SKM, ETAP, & EDSA
Power System Analysis Tutorials

Introduction to popular power system simulation software with load flow and short-circuit analysis examples

for Electrical Engineers and Technicians

Stephen P. Tubbs, P.E.
*formerly of the
Pennsylvania State University,
currently an
industrial consultant*

NOTICE TO THE READER

The author does not warrant or guarantee any of the products, equipment, or software described herein or accept liability for any damages resulting from their use.

The reader is warned that electricity and the construction of electrical equipment are dangerous. It is the responsibility of the reader to use common sense and safe electrical and mechanical practices.

AMD Athlon XP is a trademark owned by American Micro Devices, Inc.

AutoCAD is a trademark owned by Autodesk, Inc.

*CAPTOR, DAPPER, Power*Tools for Windows, PTW*, and *SKM* are trademarks owned by *SKM* Systems Analysis, Inc.

DesignBase, EDSA, and *Paladin* are trademarks owned by *EDSA* Micro Corporation.

ETAP is a trademark owned by Operation Technology, Inc.

Intel Pentium III, Intel Pentium 4, and *Intel Xeon* are trademarks owned by Intel Corporation.

Microsoft – Access, Excel, Internet Explorer, NET Framework, Server, Vista, Windows, XP Home, and *XP Professional* are trademarks owned by the Microsoft Corporation.

PSpice is a trademark owned by Cadence Design Systems, Inc.

Printed in the United States of America and United Kingdom.

ISBN   978-0-9819753-0-6

CONTENTS

iv

INTRODUCTION

Modern power system analysis is done with computers. Today, engineering firms often own several different power system analysis programs, so that they can meet different customer's requirements. Manual analysis is now only an academic pursuit, good for understanding what is happening in systems, but too slow and error prone for all but the smallest systems.

The object of this book is to teach the beginner the basics of three popular power system analysis programs. These programs are designed to simulate and analyze electrical power generation and distribution systems in normal operation and in short-circuit. The programs also have many add-on options like protection selection, arc flash analysis, transmission line sag & tension, raceway calculations, transient motor starting, etc. The programs have Demo (demonstration or trial) versions to allow people to tryout and learn about them. This book provides the engineer and technologist with information needed to use the Demo versions of *SKM, ETAP*, and *EDSA* for load flow and short-circuit analysis. The beginner learns how to use them on a small, but realistic, three-phase power system. The information gained is similar to that which students pay for in company-taught "Introduction to ..." courses. However, with this book, the student avoids paying tuition, learns at times of his own convenience, and can compare the different programs.

Of course, the programs have different strengths. For example, some have analysis capabilities that others do not have. However, they all do the same basic analyses and all three programs are good, or they would not be here. At the beginning of each program's section, information from each program's sales literature, sales staff, and web pages is presented. The information each company presents is designed to persuade the reader that its product is the best. The author makes no conclusion that one program is better than another and will neither confirm nor deny any of this sales information. The reader should contact the companies and determine for himself which product is best for his needs and budget.

In this book, load flow (power-flow) and short-circuit analyses are done on a small steady-state three-phase power system with manual methods. Then, each program is used to carry out the same analyses. Since in practice, three-phase systems are the most often analyzed, only three-phase systems will be considered in this book. The DC and single-phase capabilities of the programs will not be considered.

The person using this book should already have an analytical electrical background. Academically, he should be educated to at least the level of a university two-year electrical engineering technology program. The knowledge required for the analysis of power systems varies. If the same type of system is being analyzed over and over, then a person would not need a great deal of electrical training. Once the first simulation was done correctly, he would just be repeating the same procedure and seeing similar results. However, for unusual systems, the programs cannot be relied on to automatically produce correct results. In those cases a person with more extensive training and/or experience should interpret the programs' results.

Stephen Tubbs, P.E.

1.0 MANUAL ANALYSIS

An example system is presented that is simple enough to be manually analyzed, but complicated enough to be useful for demonstrating power system analysis programs.

This section is meant to be more of a review than a tutorial. Understanding it is necessary to understand the data input to the power system analysis programs and to understand how those programs are working. If the reader has never used the per-unit/percent system he should study a book like *Power System Analysis* by Grainger et. al. See the references in Section 1.4. If the reader has no background in *symmetrical components* and wishes to do unbalanced short-circuit analyses, he should also study a book like *Power System Analysis*.

1.1 PER-UNIT/PERCENT SYSTEM

Power systems with transformers are often analyzed with impedances, voltages, and currents expressed in the per-unit system. The major reason the per-unit system is used is to eliminate the primary side to secondary side transformer effective impedance, voltage, and current conversions that would have to be put into the calculations. Another benefit of the per-unit system is that experienced users can often quickly recognize an incorrect per-unit value. For example, power transformers usually have similar per-unit impedances, regardless of their voltage and power rating.

The percent system is practically the same thing as the per-unit system. Percent values can be converted to per-unit values by simply dividing them by 100. Many equipment manufacturers provide percent rather than per-unit values. In the example problem, some equipment specifications will be given in percent values. In the manual analysis section of the book, the percent values will be converted to per-unit values. Then calculations will be done with the per-unit values. The unit "pu" is sometimes placed after per-unit variables.

The major problems with the per-unit system are that a user may input incorrect per-unit values into his analysis or may incorrectly convert per-unit values back to actual values.

To determine per-unit values, base values in real units, i.e. Ω, volts, etc. are needed. These base values need to be carefully chosen.

Per-unit value = (Actual value)/(Base value)

1.1.1 PER-UNIT VALUES FOR A SINGLE-PHASE TRANSFORMER CIRCUIT

Figure 1-1-1-1 shows an equivalent circuit of an ideal single phase ideal transformer with a source and load.

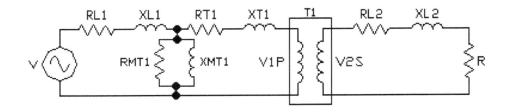

Figure 1-1-1-1 Transformer circuit with magnetizing resistance and impedance. Transformer impedances are reflected to the primary side.

To simplify power circuit calculations, the magnetizing resistance and impedance are usually ignored as shown in Figure 1-1-1-2. The elimination of magnetizing impedances causes inaccurate results with unloaded transformer circuits but is usually accurate enough for fully loaded transformers or transformers having short-circuit faults. Magnetizing impedances will not be used in this book's manual analyses.

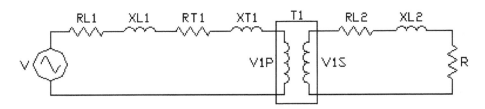

Figure 1-1-1-2 Transformer circuit without magnetizing resistance and impedance. Transformer impedances are reflected to the primary side.

The circuit of Figure 1-1-1-2 is converted to one with per-unit values.

Actual values:

V = source voltage (V rms)
V1P = transformer primary winding rated input voltage (V rms)
V1S = transformer secondary winding rated output voltage (V rms)
S1 = apparent power capability of the transformer T1 (VA)
RL1 = source and line resistance (Ω)
XL1 = source and line inductive impedance (Ω)

RT1 = transformer series resistance reflected to the primary side (Ω)
XT1 = transformer series inductive impedance reflected to the primary side (Ω)
RL2 = line resistance on the transformer secondary side (Ω)
XL2 = line inductive impedance on the transformer secondary side (Ω)
R = load resistance (Ω)

Base values are arbitrarily chosen. To make per-unit values close to one, usually they are chosen to be the same as the rated values of part of the circuit. On this circuit they are taken from the rated values of the transformer. They are:
V1P = transformer rated input voltage (V rms)
V1S = transformer rated secondary voltage (V rms)
S1 = transformer rated apparent power (VA)

With these values, other per-unit bases can be calculated.

Calculated impedance base for the primary side is:
ZBASE1P = $(V1P)^2$/S1 (Ω)

Calculated impedance base for the secondary side is:
ZBASE1S = $(V1S)^2$/S1 (Ω)

Calculated current base for the primary side is:
IBASE1P = S1/V1P (amps rms)

Calculated current base for the secondary side is:
IBASE1S = S1/V1S (amps rms)

From these the following per-unit values can be calculated:

V_{pu} = V/V1P = per-unit source voltage (pu)
$RL1_{pu}$ = RL1/ZBASE1P = source resistance (pu)
$XL1_{pu}$ = XL1/ZBASE1P = source inductive impedance (pu)
$RT1_{pu}$ = RT1/ZBASE1P = transformer series resistance (pu)
$XT1_{pu}$ = XT1/ZBASE1P = transformer series inductive impedance (pu)
$RL2_{pu}$ = RL2/ZBASE1S = line resistance on secondary side of transformer (pu)
$XL2_{pu}$ = XL2/ZBASE1S = line inductive impedance on secondary side of transformer (pu)
R_{pu} = R/ZBASE1S = load resistance (pu)

Once per-unit values have been calculated they can be incorporated into a per-unit system circuit. Notice how transformer isolation does not exist in the per-unit circuit. See Figure 1-1-1-3.

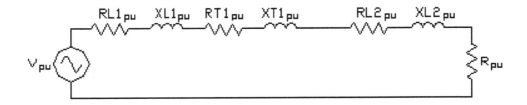

Figure 1-1-1-3 Per-unit transformer circuit.

The transformer circuit can now be solved with the simpler circuit of Figure 1-1-1-3. After solution, the per-unit values of current, voltage, and power can be returned to actual values by multiplying them by their appropriate bases.

1.1.2 PER-UNIT VALUES FOR A TWO TRANSFORMER SINGLE-PHASE SYSTEM

Often a transformer is connected to another, each of them having their own per-unit bases. A two transformer example is shown in Figure 1-1-2-1.

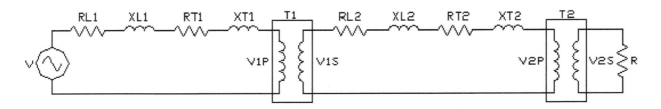

Figure 1-1-2-1 Transformer circuit with two transformers.

The values provided with Figure 1-1-2-1 are:

V = source voltage (V rms)
RL1 = source resistance (Ω)
XL1 = source inductive impedance (Ω)

V1P = transformer T1 primary winding rated input voltage (V rms)
V1S = transformer T1 secondary winding rated output voltage (V rms)
S1 = rated apparent power capability of transformer T1 (VA)

RT1pubase1 = transformer T1 series resistance reflected to the primary side (pu, using transformer T1 as its base)
XT1pubase1 = transformer T1 series inductive impedance reflected to the primary side (pu, using transformer T1 as its base)

RL2 = line resistance on the secondary side of transformer T1 (Ω)
XL2 = line inductive impedance on the secondary side of transformer T1 (Ω)

V2P = transformer T2 primary winding rated input voltage (V rms)
V2S = transformer T2 secondary winding rated output voltage (V rms)
S2 = rated apparent power capability of transformer T2 (VA)
RT2pubase2 = transformer T2 series resistance reflected to the primary side (pu, using transformer T2 as its base)
XT2pubase2 = transformer T2 series inductive impedance reflected to the primary side (pu, using transformer T2 as its base)
R = load resistance (Ω)

The circuit of Figure 1-1-2-1 is converted to one with per-unit values. See Figure 1-1-2-2.

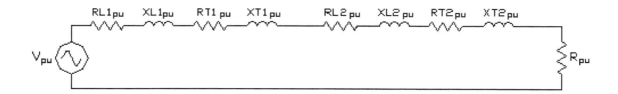

Figure 1-1-2-2 Transformer circuit with two transformers. Values are in the per-unit system with transformer T1 as the base.

The transformer T1 will be selected as the base for determining per-unit values throughout the circuit. As in the one transformer example in Section 1.1.1, the base values are:
V1P = transformer T1 primary winding rated input voltage (V rms)
V1S = transformer T1 secondary winding rated output voltage (V rms)
S1 = transformer T1 rated apparent power (VA)

With these values, other per-unit bases can be calculated.

Calculated impedance base for the transformer T1 primary side is:
ZBASE1P = $(V1P)^2/S1$ (Ω)

Calculated impedance base for the transformer T1 secondary side is:
ZBASE1S = $(V1S)^2/S1$ (Ω)

With these bases, per-unit values can be found for the impedances and voltage source in series with the input and output of the transformer T1. As with the single transformer example of Section 1.1.1:

V_{pu} = V/V1P = per-unit source voltage (pu)
$RL1_{pu}$ = RL1/ZBASE1P = source resistance (pu)
$XL1_{pu}$ = XL1/ZBASE1P = source inductive impedance (pu)
$RT1_{pu}$ = Transformer T1 per-unit resistance given with the transformer T1 data
$XT1_{pu}$ = Transformer T1 per-unit inductive impedance given with the transformer T1 data
$RL2_{pu}$ = RL2/ZBASE1S = line resistance on secondary side of transformer T1 (pu)
$XL2_{pu}$ = XL2/ZBASE1S = line inductive impedance on secondary side of transformer T1 (pu)

Per-unit values were given for the transformer T2, but these values were calculated relative to the transformer T2 base. If the transformers T1 and T2 do not have the same bases, then the T2-based per-unit values given for the transformer T2 should be converted to T1-based per-unit values.

Impedance base used by the transformer T2 per-unit values transformer primary side is:
ZBASE2P = $(V2P)^2$/S2 (Ω)

Calculating the transformer T2 primary side resistance to the base of the transformer T1:
$RT2_{pu}$ = RT2pubase1 = (RT2pubase2) (ZBASE2P)/(ZBASE1S)
$\quad$ = (RT2pubase2) $(V2P^2$/S2)/$(V1S^2$/S1) = (RT2pubase2) $(V2P/V1S)^2$ (S1/S2)

If V2P = V1S then $RT2_{pu}$ = (RT2pubase2) (S1/S2)

Calculating the transformer T2 primary side inductive impedance to the base of the transformer T1:
$XT2_{pu}$ = XT2pubase1 = (XT2pubase2) (ZBASE2P)/(ZBASE1S)
$\quad$ = (XT2pubase2) $(V2P^2$/S2)/$(V1S^2$/S1) = (XT2pubase2) $(V2P/V1S)^2$ (S1/S2)

If V1S = V2P then $XT2_{pu}$ = (XT2pubase2) (S1/S2)

The value for the load resistor, R, also needs to be converted to the per-unit base of the transformer T1.

R_{pu} = R $(V2P/V2S)^2$/ZBASE1S = R $(V2P/V2S)^2$/$(V1S^2$/S1)

If V1S = V2P then R_{pu} = R S1/$V2S^2$

After the circuit is solved with per-unit values, the per-unit currents and voltages will be multiplied by their respective base values to get actual values.

1.1.3 PER-UNIT VALUES FOR A BALANCED THREE-PHASE TRANSFORMER CIRCUIT

Balanced three-phase systems are easier to solve than unbalanced ones. Balanced three-phase systems can be solved by using a single-phase equivalent circuit of one phase of the three-phase system. Any balanced three-phase system can be converted to a single-phase equivalent circuit for analysis purposes. However, the Y-Y three-phase transformer circuit is discussed here, since it is the easiest to visualize as its single-phase equivalent. Figure 1-1-3-1 shows an example of a Y source supplying a Y load via Y-Y connected transformers.

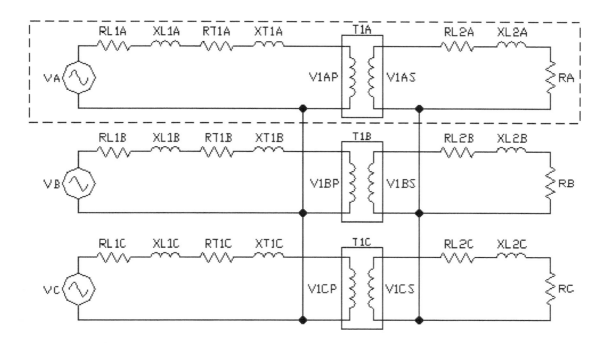

Figure 1-1-3-1   Y source to Y-Y transformers to Y load circuit showing all three phases of a balanced three-phase circuit. The portion of the circuit enclosed in a dashed rectangle, the A-phase, can be solved in the same way as the single-phase circuit of Section 1.1.1.

Per-unit values of one phase of a balanced Y-Y circuit are found as they were in Section 1.1.1.

Balanced Y-Δ, Δ-Y, and Δ-Δ transformer circuits can be converted to Y-Y equivalents and then per-unit values found for the equivalents. Note that a 30° phase shift occurs with Y-Δ and Δ-Y transformer circuits. The magnitude calculations will be correct when Y-Δ and Δ-Y transformer circuits are represented as Y-Y equivalents, but the phase shift will be different by 30°.

1.2 SYMMETRICAL COMPONENTS

Symmetrical components is a method of solving unbalanced poly-phase circuits. It is used in manual calculations and in power system analysis programs. The method was developed by Dr. C. L. Fortesque of the Westinghouse Corporation. He presented it in an AIEE (American Institute of Electrical Engineers) paper in 1918.

Usually *symmetrical components* are used to analyze balanced three-phase systems that are experiencing unbalanced short circuits. With extra effort and certain complications, *symmetrical components* can also be used on unbalanced poly-phase circuits. However, this usually is not done.

When used to analyze unbalanced three-phase short circuits, a three-phase circuit is converted into three single-phase sequence circuits. The single-phase circuits are called the positive sequence circuit, negative sequence circuit, and zero sequence circuit. These circuits are then combined in different configurations to produce results applicable to line-to-line, line-to-neutral, and line-to-line-to-neutral short-circuit conditions.

Whole books and chapters of books have been written about the *symmetrical components* method. If the reader will be doing only load flow and balanced short-circuit analysis, where the three-phase system is balanced, then *symmetrical components* could be skipped. If three-phase unbalanced short-circuit analysis is to be done, then the reader should learn the basics of *symmetrical components*. See Section 1.4 for *symmetrical components* references.

In the following example the impedance suffix 1 means positive sequence, 2 negative sequence, and 0 zero sequence. For instance, for cable L1, $RL11_{pu}$ is the per-unit positive sequence resistance, $RL12_{pu}$ is the per-unit negative sequence resistance, and $RL10_{pu}$ is the per-unit zero sequence resistance.

1.3 EXAMPLE SYSTEM ANALYZED MANUALLY

An example is presented that manually analyzes the system of Figure 1-3-1 and the equipment data that follows it to achieve the following goals:

a) Do a load flow analysis of the system during normal balanced full-load operation. In this determine the main bus voltage and the currents through circuit-breaker C1, circuit-breaker C2, and fuse F1.
b) Do a short-circuit analysis of the system with balanced (bolted-fault) short circuits occurring at the outputs of C1, C2, and F1.
c) Do a short-circuit analysis of the system with an unbalanced line-to-neutral short-circuit fault occurring at the motor input. In this determine the maximum short-circuit current through the circuit breaker C1.

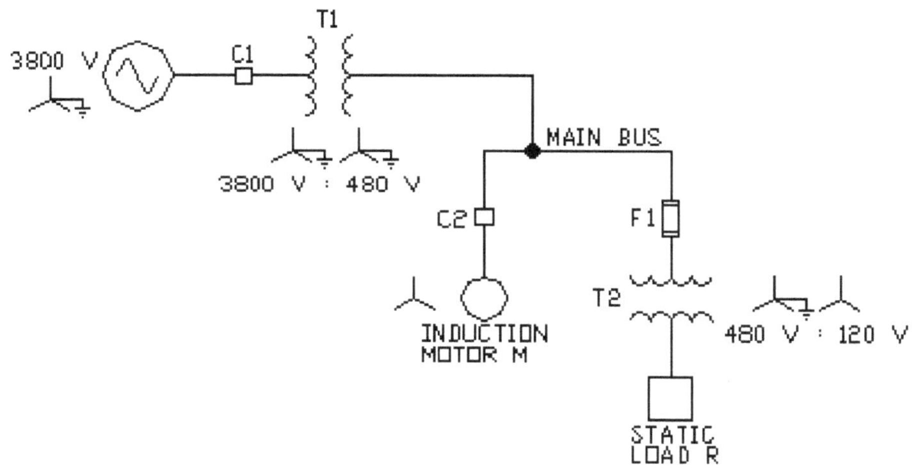

Figure 1-3-1 One-line diagram of the example system.

The solution of this problem requires the gathering of component data, the conversion of the data to one per-unit base, entry of the per-unit data into equivalent circuit diagrams, solution of the circuit diagrams, and conversion of the solution per-unit data back to actual values.

1.3.1 COMPONENT DATA

There are many component values needed to analyze the system. These values are manufacturer's equipment data, lengths of conductors, etc. This section will convert raw equipment data and local base percent data to common base per-unit values that can be used in per-unit equivalent circuits. Negative and zero sequence values apply only to *symmetrical components* calculations. Positive sequence values are used in both balanced steady-state calculations and *symmetrical components* calculations.

The base for per-unit calculations will be the transformer T1 rated voltages and power.

Note that care needs to be taken to not confuse line-to-line voltages and total power with line-to-neutral voltages and phase power. To simplify calculations, most three-phase calculations will use line-to-neutral voltages and phase powers (1/3 of total power) values.

1.3.1.1 Transformers

1.3.1.1.1 Transformer T1

Transformer T1 will be used as the base for per-unit values throughout the system.

Supplied Raw Data
> 3 phase
> 60 Hz
> Y-Y connection with solidly grounded neutrals
> Rated power capacity .6 MVA
> Primary rated voltage 3.8 kV line-to-line
> Secondary rated voltage .48 kV line-to-line
> Positive sequence %RT11 = 1.73%
> Positive sequence %XT11 = 1.52%
> Negative sequence %RT12 = 1.73%
> Negative sequence %XT12 = 1.52%
> Zero sequence %RT10 = 1.73%
> Zero sequence %XT10 = 1.52%

Calculated Per-Unit Impedances
> $RT11_{pu} = 1.73/100 = .0173$
> $XT11_{pu} = 1.52/100 = .0152$
> $RT12_{pu} = 1.73/100 = .0173$
> $XT12_{pu} = 1.52/100 = .0152$
> $RT10_{pu} = 1.73/100 = .0173$
> $XT10_{pu} = 1.52/100 = .0152$

Base Values
> Primary side line-to-line voltage = 3800 V rms
> Primary side line-to-neutral voltage = $V1P = 3800/(3)^{1/2} = 2194$ V rms
> Secondary side line-to-line voltage = 480 V rms
> Secondary side line-to-neutral voltage = $V1S = 480/(3)^{1/2} = 277.1$ V rms
> Rated apparent power per phase = S1 = 600000/3 = 200000 VA
> Primary side line current = 200000/2194 = 91.16 A rms
> Secondary side line current = 200000/277.1 = 721.7 A rms
> Primary side impedance = $ZBASE1P = (2194)^2/200000 = 24.07\ \Omega$
> Secondary side impedance = $ZBASE1S = (277.1)^2/200000 = .3839\ \Omega$

1.3.1.1.2 Transformer T2

Supplied Raw Data

　　3 phase
　　60 Hz
　　Y-Y connection with the primary solidly grounded and the secondary neutral open
　　Rated power capacity = .090 MVA
　　Primary rated voltage 480 V rms line-to-line
　　Secondary rated voltage 120 V rms line-to-line
　　Positive sequence (base transformer T2) %RT21 = 2.0%
　　Positive sequence (base transformer T2) %XT21 = 1.8%
　　Negative sequence (base transformer T2) %RT22 = 2.0%
　　Negative sequence (base transformer T2) %XT22 = 1.8%
　　Zero sequence (base transformer T2) %RT20 = 2.0%
　　Zero sequence (base transformer T2) %XT20 = 1.8%

Base Values

　　Primary side line-to-line voltage = 480 V rms
　　Primary side line-to-neutral voltage = V2P = $480/(3)^{1/2}$ = 277.1 V rms line-to-neutral
　　Secondary side line-to-line voltage = 120 V rms
　　Secondary side line-to-neutral voltage = V2S = $120/(3)^{1/2}$ = 69.3 V rms line-to-neutral
　　Rated apparent power per phase = S2 = 90000/3 = 30000 VA

Calculated Per-Unit (Base Transformer T1) Impedances (see Method in Section 1.1.2)

　　$RT21_{pu}$ = (2.0/100) (200000/30000) = .1333
　　$XT21_{pu}$ = (1.52/100) (200000/30000) = .1013
　　$RT22_{pu}$ = (2.0/100) (200000/30000) = .1333
　　$XT22_{pu}$ = (1.52/100) (200000/30000) = .1013
　　$RT20_{pu}$ = (2.0/100) (200000/30000) = .1333
　　$XT20_{pu}$ = (1.52/100) (200000/30000) = .1013

1.3.1.2 Source

1.3.1.2.1 Source voltage

Assumed Phase Angle

　　The phase angle of the A-phase line-to-neutral source voltage (the phase used for calculations) is assumed to be 0 degrees.

Supplied Raw Data
   3 phase
   60 Hz
   Y connection with a solidly grounded neutral
   3.8 kV line-to-line
   ABC sequence

Base Values
   Output line-to-line voltage = 3800 V rms
   Output line-to-neutral voltage = $3800/(3)^{1/2}$ = 2194 V rms line-to-neutral

Calculated Per-Unit (Base Transformer T1) Voltage Value
   V_{pu} = 3800/3800 = 1.000

1.3.1.2.2 Source impedances

   The source impedances include the line impedances to the transformer T1.

   Only steady-state impedance is used in these calculations. Short-term transient circuit operation, which would require the use of transient and sub-transient impedances, will not be done here.

Supplied Raw Data
   60 Hz
   Positive sequence RS1 = .2 Ω
   Positive sequence XS1 = .3 Ω
   Negative sequence RS2 = .2 Ω
   Negative sequence XS2 = .3 Ω
   Zero sequence RS0 = .1 Ω
   Zero sequence XS0 = .15 Ω

Calculated Per-Unit (Base Transformer T1) Impedances (see Calculation Method in Section 1.1.2)
   $RS1_{pu}$ = .2/ZBASE1P = .008310
   $XS1_{pu}$ = .3/ZBASE1P = .01246
   $RS2_{pu}$ = .2/ZBASE1P = .008310
   $XS2_{pu}$ = .3/ZBASE1P = .01246
   $RS0_{pu}$ = .1/ZBASE1P = .004155
   $XS0_{pu}$ = .15/ZBASE1P = .006232

1.3.1.3 Line and Bus Impedances

1.3.1.3.1 Cables from source to circuit breaker C1

These impedances are included with the source impedances. See Section 1.3.1.2.2.

1.3.1.3.2 Conductors from circuit breaker C1 to transformer T1, bus bars from transformer T1 to MAIN BUS, bus bars from MAIN BUS to circuit breaker C2, bus bars from MAIN BUS to fuse F1, and bus bars from fuse F1 to transformer T2

For all of these, the distances are short and the conductors large in cross-sectional area, so their impedances are assumed to be zero.

1.3.1.3.3 Cable L1 from circuit breaker C2 to motor M

Supplied Raw Data
Three-phase four wire
200 feet
AWG # 2/0 (.000078 Ω/ft) used for all four wires
Inductive impedance per foot = .00003Ω/ft. (closely spaced conductors)

Calculated Per-Unit (Base Transformer T1) Impedances
$RL11_{pu}$ = (.000078) (200)/ZBASE1S = .04064
$XL11_{pu}$ = (.00003) (200)/ZBASE1S = .01563
$RL12_{pu}$ = (.000078) (200)/ZBASE1S = .04064
$XL12_{pu}$ = (.00003) (200)/ZBASE1S = .01563
$RL10_{pu}$ = (.000078) (200)/ZBASE1S = .04064
$XL10_{pu}$ = (.00003) (200)/ZBASE1S = .01563

1.3.1.3.4 Cable L2 from transformer T2 to static load R

Supplied Raw Data
Three-phase four wire
200 feet
AWG # 1/0 (.000098 Ω/ft) used for all four wires
Inductive impedance per foot = .00003Ω/ft. (closely spaced conductors)

Calculated Per-Unit (Base Transformer T1) Impedances

$RL21_{pu} = (.000098) (200) (277.1/69.28)^2/ZBASE1S = .8169$
$XL21_{pu} = (.00003) (200) (277.1/69.28)^2/ZBASE1S = .2501$
$RL22_{pu} = (.000098) (200) (277.1/69.28)^2/ZBASE1S = .8169$
$XL22_{pu} = (.00003) (200) (277.1/69.28)^2/ZBASE1S = .2501$
$RL20_{pu} = (.000098) (200) (277.1/69.28)^2/ZBASE1S = .8169$
$XL20_{pu} = (.00003) (200) (277.1/69.28)^2/ZBASE1S = .2501$

1.3.1.4 Motor M

An induction motor in steady-state operation can be represented by an inductance in series with a resistance. An accurate steady-state motor model would adjust a motor's effective impedance as applied voltage and motor speed vary. A computer program using iterative methods would probably be used to do the required calculations.

Here, to simplify the manual analysis, the induction motor will be represented by a fixed inductance and resistance. The values of these will be calculated from the motor's steady-state full-load power, rated voltage, rated power factor, and rated efficiency. The use of fixed inductance and resistance values will result in some inaccuracy when the motor supply voltage is not at its rated value. However, in this example problem, the results are sufficient.

Supplied Raw Data
3 phase
480 V rms line-to-line
250 Hp
Y-Y connection with an isolated neutral
Service factor 1.0
92.86% efficiency at full-load
91.8% power factor at full-load

Calculated Y Equivalent Values at Full-load
Current into the motor at full-load:
$$I = [(250) (746)]/[(.9286) (.918) (3)^{1/2} (480)] = 263.1 \text{ A rms}$$

Effective positive sequence motor resistance per phase
$$RM1 = [(480) (.918)]/[(3)^{1/2} (263.1)] = .9670 \ \Omega$$

Effective positive sequence motor inductive impedance per phase
$$XM1 = (.9670) \{\tan[\arccos(.918)]\} = .4177 \ \Omega$$

Calculated Per-Unit (Base Transformer T1) Y Equivalent Impedances at Full-Load

Effective motor resistance per phase
$RM1_{pu} = .9670/ZBASE1S = 2.519$
$RM2_{pu} \sim RM1_{pu}/10 = 2.519/10 = .2519$ <Note the 1/10 factor is a rough estimate for converting positive to negative sequence induction motor impedance. It assumes that the motor is operating with close to 5% slip. See Wagner and Evans book, reference 2 of Section 1.4, for a more detailed explanation.>

Effective motor inductive impedance per phase
$XM1_{pu} = .4177/ZBASE1S = 1.088$
$XM2_{pu} = XM1_{pu} = 1.088$

Since the motor has an isolated neutral the $RM0_{pu}$ and $XM0_{pu}$ can both be assumed to be infinite.

1.3.1.5 Static Load R

Supplied Raw Data
3 phase
120 V rms line-to-line
30 kW
Y connection with ungrounded neutral
100% power factor

Base Values
Input line-to-line voltage = 120 V rms
Input line-to-neutral voltage = $120/(3)^{1/2} = 69.28$ V rms line-to-neutral
Rated apparent power per phase 30000/3 = 10000 VA

Calculated Y Equivalent Resistance Value

$R = (69.28)^2/10000 = .4800\ \Omega$

Calculated Per-Unit (Base Transformer T1) Y Equivalent Impedances

$R1_{pu} = .4800(277.1/69.28)^2/ZBASE1S = 20.01$
$R2_{pu} = .4800(277.1/69.28)^2/ZBASE1S = 20.01$
$R0_{pu} = $ infinity, open circuit

1.3.2 LOAD FLOW ANALYSIS

Load flow analysis is the determination of steady-state voltages, currents, and powers in the balanced system. Figure 1-3-2-1 shows one phase of the system. This circuit is the same as the positive sequence circuit that is used later in *symmetrical components* analysis. In the circuit values' subscripts, the "1"s to the left of the "pu"s indicate positive sequence values. Unless otherwise stated, voltages and currents refer to the A phase.

Phasor values will be represented by larger type bold characters. For example, $\mathbf{V_{pu}}$ is a phasor voltage that includes the voltage magnitude of V_{pu} and the angle of V_{pu}.

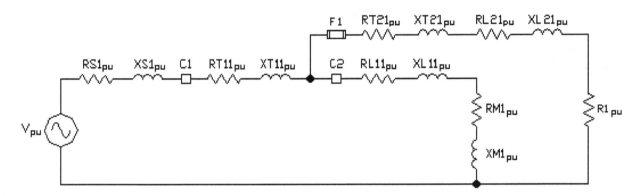

Figure 1-3-2-1 One phase of the example problem. This is also the *symmetrical components* positive sequence diagram.

1.3.2.1 Determining the Main Bus Voltage

Phasor values $\mathbf{A}$, $\mathbf{B}$, $\mathbf{C}$, and $\mathbf{D}$ are used to simplify calculations.

$\mathbf{A} = RS1_{pu} + XS1_{pu}j + RT11_{pu} + XT11_{pu}j$

$\quad = .008310 + .01246j + .0173 + .0152j = .03770 \angle 47.20°$

$\mathbf{B} = RT21_{pu} + XT21_{pu}j + RL21_{pu} + XL21_{pu}j + R1_{pu}$

$\quad = .1333 + .1013j + .8169 + .2501j + 20.01 = 20.96 \angle .9625°$

$\mathbf{C} = RL11_{pu} + XL11_{pu}j + RM1_{pu} + XM1_{pu}j$

$\quad = .04064 + .01563j + 2.519 + 1.088j = 2.787 \angle 23.32°$

$$\mathbf{D} = (\mathbf{B}^{-1} + \mathbf{C}^{-1})^{-1}$$

$$= [(20.96 \angle .9625^{\circ})^{-1} + (2.787 \angle 23.32^{\circ})^{-1}]^{-1} = 2.480 \angle 20.74^{\circ}$$

Per unit voltage at the main bus is:

$$\mathbf{VBUS_{pu}} = \mathbf{V_{pu}} \, [\mathbf{D}/(\mathbf{A} + \mathbf{D})]$$

$$= (1.000 \angle 0^{\circ}) \, [2.480 \angle 20.74^{\circ} / (.03770 \angle 47.20^{\circ} + 2.480 \angle 20.74^{\circ})]$$

$$= .9865 \angle -.38^{\circ}$$

$$\mathbf{VBUS_{actualLL}} = 480(\mathbf{VBUS_{pu}}) = 480(.9865 \angle -.38^{\circ}) = \underline{473.5 \angle -.38^{\circ}} \, \text{V rms line-to-line}$$

1.3.2.2 Determining Current through the Circuit Breaker C2

$$\mathbf{IC2_{pu}} = \mathbf{VBUS_{pu}}/\mathbf{C}$$

$$= .9865 \angle -.38^{\circ}/2.787 \angle 23.32^{\circ} = .3540 \angle -23.70^{\circ}$$

$$\mathbf{IC2_{actual}} = 721.7(\mathbf{IC2_{pu}}) = 721.7(.3540 \angle -23.70^{\circ}) = \underline{255.5 \angle -23.70^{\circ}} \, \text{A rms}$$

1.3.2.3 Determining current through the fuse F1

$$\mathbf{IF1_{pu}} = \mathbf{VBUS_{pu}}/\mathbf{B}$$

$$= .9865 \angle -.38^{\circ}/20.96 \angle .9625^{\circ} = .04707 \angle -1.342^{\circ}$$

$$\mathbf{IF1_{actual}} = 721.7(\mathbf{IF1_{pu}}) = 721.7(.04707 \angle -1.342^{\circ}) = \underline{33.97 \angle -1.342^{\circ}} \, \text{A rms}$$

1.3.2.4 Determining current through the circuit breaker C1

$$\mathbf{IC1_{actual}} = (\mathbf{IC2_{actual}} + \mathbf{IF1_{actual}})(277.1/2194)$$

$$= (255.5 \angle -23.70^{\circ} + 33.97 \angle -1.342^{\circ}) \, (277.1/2194) = \underline{36.30 \angle -21.12^{\circ}} \, \text{A rms}$$

1.3.3 BALANCED SHORT-CIRCUIT ANALYSIS

Balanced line-to-line short circuits are not as common as unbalanced short circuits. However, they are easier to analyze. Their relative ease of analysis has the benefit of reducing the likelihood of calculation errors. Current values found for balanced line-to-line-to-line short circuits are usually close to worse case current values of unbalanced short circuits.

Breakers and fuses are rated by their manufacturers to properly operate to stated values of balanced overcurrent.

The major supplier of short-circuit current is usually the supply voltage. However, for a few cycles, a motor will also contribute current to a short circuit. The magnitude of the current an induction motor contributes is close to its initial starting (locked rotor) current. Induction motor starting current is about five times that of its full-load current.

Note that the phase angle of V_{pu} and the circuit's currents given below are for the A phase.

1.3.3.1   Current through the Circuit Breaker C1 when a Balanced Short Circuit Occurs at the C1 Output

Determining the source contributed short-circuit current:

The drawing of Figure 1-3-2-1 is redrawn with a balanced short circuit at the C1 output. See Figure 1-3-3-1-1.

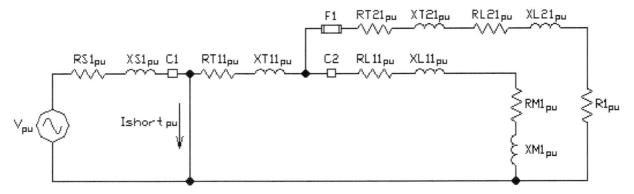

Figure 1-3-3-1-1 One phase of a balanced short circuit at the output of the circuit breaker C1.

The circuit can be simplified. See Figure 1-3-3-1-2.

Figure 1-3-3-1-2 Simplified circuit of one phase of a balanced short circuit at the output of the circuit breaker C1.

$$\mathbf{Ishort_{pu}} = \mathbf{V_{pu}}/(RS1_{pu} + XS1_{pu}j)$$

$$= 1.000\angle 0^{o}/(.01498\angle 56.29^{o}) = 66.77\angle -56.29^{o}$$

Source contributed short-circuit current =
$$\mathbf{Ishort_{actual}} = 91.16(\mathbf{Ishort_{pu}}) = 91.16(66.77\angle -56.29^{o}) = \underline{6088\angle -56.29^{o} \text{ A rms}}$$

Determining the motor contributed short-circuit current:

From Section 1.3.1.4, motor full-load current = 263.1 A rms

Motor contribution to the short circuit as seen at the motor terminals = 5 x 263.1 = 1316 A rms

The impedance for motor contributed short-circuit current is much lower in the path through the short circuit than in the path through the static load. It is reasonable to make the assumption that all the motor generated current goes to the short circuit.

The actual motor contributed fault current seen at the short circuit is reduced by the transformer ratio.

Motor contributed short-circuit current = (1316) (277.1/2194) = 166.2 A rms

1.3.3.2 Currents through the Circuit Breaker C2 or Fuse F1 when a Balanced Short Circuit Occurs at the Output of the Circuit Breaker C2 or Fuse F1

Determining the source generated short-circuit current:

Since the resistances of the circuit breaker C2, fuse F1, and the wiring back to the bus are practically zero the available short-circuit currents at the outputs of C2 and F1 have the practically the same values. The drawing of Figure 1-3-2-1 is redrawn with a short circuit at the C2 output. See Figure 1-3-3-2-1.

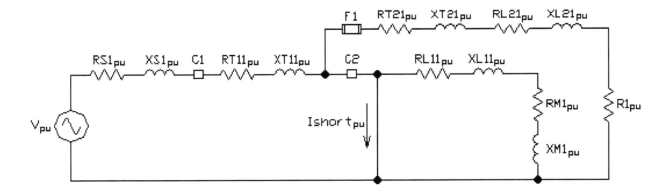

Figure 1-3-3-2-1 One phase of a balanced short circuit at the bus.

The circuit can be simplified. See Figure 1-3-3-2-2.

Figure 1-3-3-2-2 Simplified circuit of one phase of a balanced short circuit at the main bus.

Ishort$_{pu}$ = V$_{pu}$/A

$$= 1.000\angle 0°/(.03770\angle 47.20°) = 26.53\angle -47.20°$$

Source contributed short-circuit current =
Ishort$_{actual}$ = 721.7(**Ishort$_{pu}$**) = 721.7(26.53$\angle$-47.20°) = <u>19146$\angle$-47.20° A rms</u>

Determining the motor generated short-circuit current:

From Section 1.3.1.4, motor full-load current = 263.1 A rms

Approximate motor contribution to the short circuit at the motor terminals = 5 x 263.1 = <u>1316 A rms</u>

1.3.4 UNBALANCED SHORT-CIRCUIT ANALYSIS, LINE-TO-GROUND SHORT CIRCUIT AT THE MOTOR INPUT

To determine current to ground during an A-phase line-to-ground short circuit, the *symmetrical components* positive, negative, and zero sequence circuits are connected in series. The combined circuit is solved for a current that is then converted to the actual expected line-to-ground short-circuit current.

The positive sequence circuit of Figure 1-3-2-1 will be combined with the negative sequence circuit of Figure 1-3-4-1 and zero sequence circuit of Figure 1-3-4-2. The result is the *symmetrical components* circuit of Figure 1-3-4-3. This represents an A-phase to ground short circuit at the motor input.

Since the motor does not have a connected neutral, it will not contribute to line-to-ground short-circuit current.

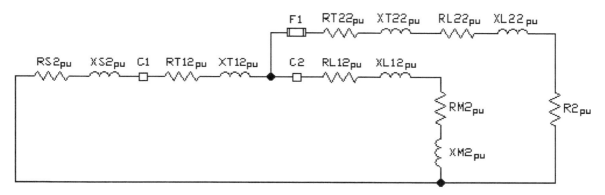

Figure 1-3-4-1 Negative sequence circuit of the example problem.

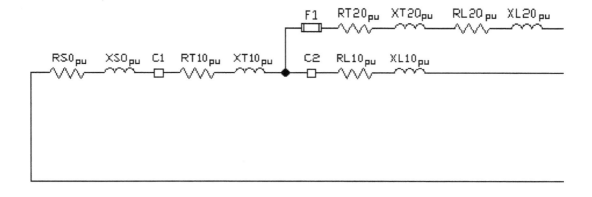

Figure 1-3-4-2 Zero sequence circuit of the example problem.

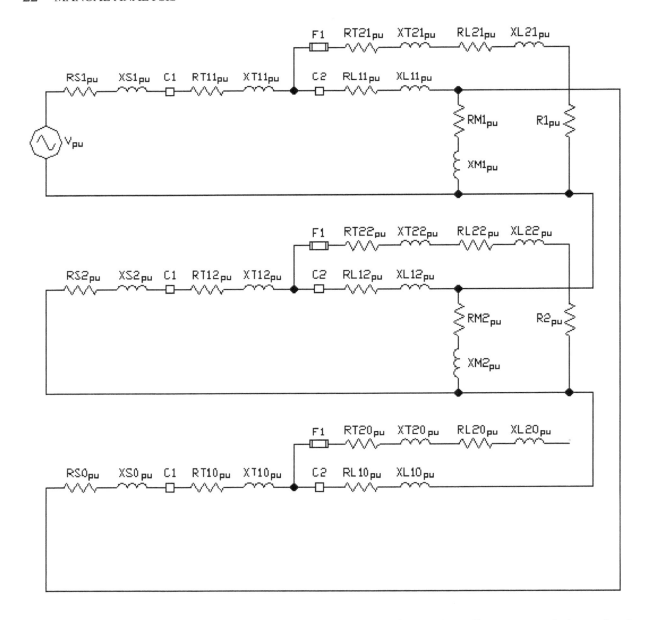

Figure 1-3-4-3 *Symmetrical components* circuit for solving for A-phase line to ground short-circuit current.

Calculations can be simplified by removing the F1 and motor (RM1 and RM2) legs of Figure 1-3-4-3. This is justified since the impedances of the F1 and M legs are much larger than their parallel impedances. Figure 1-3-4-4 shows the resultant circuit.

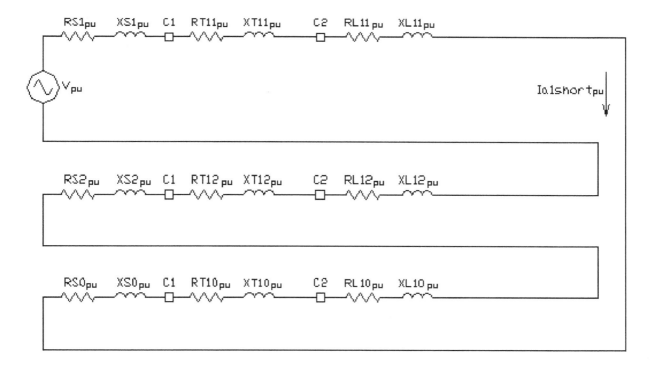

Figure 1-3-4-4 Simplified *symmetrical components* circuit for solving for A-phase line-to-ground short-circuit current.

The A-phase line-to-ground current is found with the following equations:

Iashort$_{pu}$ = 3Ia1short$_{pu}$

$$= 3[V_{pu}/(RS1_{pu} + XS1_{pu}j + RT11_{pu} + XT11_{pu}j + RL11_{pu} + XL11_{pu}j +$$

$$RS2_{pu} + XS2_{pu}j + RT12_{pu} + XT12_{pu}j + RL12_{pu} + XL12_{pu}j +$$

$$RS0_{pu} + XS0_{pu}j + RT10_{pu} + XT10_{pu}j + RL10_{pu} + XL10_{pu}j)]$$

$$= 3[1.000\angle 0° /(.00831 + .01246j + .0173 + .0152j + .04064 + .01563j +$$

$$.00831 + .01246j + .0173 + .0152j + .04064 + .01563j +$$

$$.004155 + .006232j + .0173 + .0152j + .04064 + .01563j)]$$

$$= 3[1.000\angle 0° /(.1946 + .1236j)] = 13.01 \angle -32.42°$$

Iashort$_{actual}$ = 721.7(**Iashort$_{pu}$**) = 721.7(13.01 $\angle$ -32.42°) = 9389 $\angle$ -32.42° A rms

1.4 MANUAL CIRCUIT ANALYSIS REFERENCES

1) Grainger, John and Stevenson, William D., *Power Systems Analysis (Power & Energy)*, McGraw-Hill Inc., 1994. It is 784 pages long and 6" x 9". The hard cover edition lists for $151.80 and the paperback for $78.75. This text covers a wide range of electrical power system analysis methods that would be of greatest interest to those in the electrical power industry, but also may be useful to others doing electrical power circuit calculations. It has material on the per-unit system and *symmetrical components*. It is a very good book.

2) Wagner, C. F. and Evans, R. D., *Symmetrical Components*, original edition 1933, reprint edition 1982, original edition McGraw-Hill, reprint edition Robert E. Krieger Publishing Co., Malabar, Florida. It is 437 pages long and 6" x 8". Used copies were found on the internet for $40.50. This is a good book.

3) Cooper Bussman, Part 3 Component Protection for Electrical Systems, Bulletin EDP-3. Currently available for free on http://www.bussman.com/library/docs/Edp-3.pdf This provides practical basic information on fuse and circuit-breaker selection.

4) Nochumson, Charles J., Selective Coordination, Eaton Electrical Inc., 2006. Currently, available for free on a web page that can be accessed through http://www.eaton.com/EatonCom/Markets/Electrical/Products/CircuitProtection/index.htm This provides practical information on circuit-breaker and fuse coordination.

5) National Fire Protection Association, Inc., NEC *2008 NFPA 70: National Electrical Code*, 2008. It is 864 pages long and 8.5" x 11". Its paperback list price is $75.00. The *National Electrical Code* has been adopted by many U.S. regional governments and industries as the minimum required standard for electrical construction. Anybody doing electrical work for the U.S. public should have and use it.

2.0 *SKM*

SKM Systems Analysis, Inc. is a California-based corporation founded in 1972 by three engineers with a desire to automate electrical design calculations. *SKM* is an acronym for their initials.

They originally developed the *DAPPER* (Distribution Analysis for Power Planning Evaluation and Reporting) power system analysis program on a mainframe at the UCLA Computer Center. By 1978, *DAPPER* was running in time-share mode on the Control Data Corporation mainframe and being used by engineers in the United States and internationally.

The first PC version of *DAPPER* was released in 1981, followed in 1983 by the *CAPTOR* (Computer Aided Plotting for Time Overcurrent Reporting). *A_FAULT* was developed in 1988 and *IEC_FAULT* was developed in 1991, to provide ANSI and IEC-909 fault calculations, respectively. The dynamic simulation program *I*SIM* and the harmonic analysis program *HI_WAVE* filled out the *Microsoft DOS*-based product line in 1992.

In 1995, *SKM* released *PTW* (*Power*Tools for Windows*) - completely built from ground up using the latest development tools. The *PTW* line of programs offers the same studies that have been available in *DOS*, but adds greater flexibility, enhanced ease of use, and graphical functions not available in *DOS*. *PTW* was first released in December 1995 with the base *DAPPER* and *CAPTOR* modules. *A_FAULT*, *IEC_FAULT* and *TMS* were added the following year. *HI_WAVE* was added in 1998, I*SIM in 1999, and the Device Evaluation module in 2000. The Ground Grid, Reliability and Single-Phase/Unbalanced 3-Phase studies were all added in 2001.

2.1 FEATURES, AS STATED BY *SKM*

Statement from the *SKM* sales staff:

"Why Choose *SKM*?

Quite simply, *Power*Tools for Windows* is the Gold Standard to which all other software packages must be compared. No other software can match *SKM Power*Tools* for project analysis flexibility, speed, or database integration. And with over 35,000 users worldwide, no other software package has the breadth of support from the professionals in the engineering community of *SKM Power*Tools for Windows*. *SKM* software is chosen by 39 of the top 40 Electrical Engineering firms in the world. From industry leaders like Cutler-Hammer, GE, Square-D and Siemens Westinghouse, to independent contractors and leaders in every industry, *Power*Tools* is the application of choice when it comes to electrical engineering software.

However, there is much more to *Power*Tools* than industry wide acceptance of a superior product. *SKM Power*Tools for Windows* is backed by a world class technical support staff that is comprised entirely of electrical engineers. With over 20 percent of our support staff holding registered professional engineer certifications, you can be assured that you are talking to someone who understands your problems and not just someone who is reading answers to you from a handbook. In addition to their degrees and certifications, *SKM* support engineers are required to have at least three full years of experience with the *Power*Tools* suite of software and to maintain their knowledge by instructing classes on the various *Power*Tools* Study Modules.

If top notch products and unparalleled customer service aren't enough, *SKM* Systems Analysis, Inc. offers even more. Your purchase is backed by a company with 35 years in the industry. We were the first to market, and have consistently offered our clients the best electrical engineering product available. Couple this with our industry leading training classes and our customizable Corporate Training Program and you can see why *SKM* offers each and every customer a true advantage!"

The 2008 *SKM* Power*Tools Electrical Engineering Software sales brochure states:

"*SKM* Power*Tools software helps you design and analyze electrical power systems. Interactive graphics, rigorous calculations and a powerful database efficiently organize, process and display information."

SKM advertising brochures describe *SKM PTW* program modules:

"1) *PTW DAPPER* is an integrated set of modules for Three-Phase Power System Design and Analysis including rigorous load flow and voltage drop calculations, impact motor starting, traditional fault analysis, demand and design load analysis, feeder raceway and transformer sizing, and panel, MCC, and switchboard schedule specification.

2) *PTW CAPTOR* produces time versus current coordination drawings with one-line diagrams and setting reports. It lets you coordinate protective devices with interactive on-screen graphics, and provides a comprehensive library. You can print on preprinted graph paper or on plain paper with custom grids and layouts.

3) *PTW Arc Flash* allows you to build a system model for your entire power system. Once the model is built, the Arc Flash module <see 4) below> calculates the incident energy and flash boundary at every location in the system.

4) *ArcCalc* calculates the incident energy and arc flash boundary and selects the Personal Protection Equipment (PPE) for a single point in a simplified power system. Minimum and maximum arcing short-circuit currents are calculated using broad tolerances to provide conservative results with estimated system data. Incident energy, arc flash boundaries and PPW are calculated following the NFPA 70E and IEEE 1584 standards.

5) *PTW A_FAULT* provides fault calculations in full compliance with ANSI C37 standards. It offers separate solutions for low, medium, and high voltage systems and for symmetrical, momentary, and interrupting calculations as defined in the standards.

6) *PTW IEC_60909* calculates short-circuit currents using the equivalent voltage source as required by the IEC 60909 standard. With *PTW* IEC_60909, three-phase and unbalanced fault duties for electrical power systems are calculated in compliance with the IEC 60909 standards for low, medium, and high voltage systems.

7) *PTW IEC_61363 Short-Circuit Study* module calculates the current that flows in an electrical power system under abnormal conditions. These currents must be calculated in order to adequately specify electrical apparatus withstand and interrupt ratings and selectively coordinate time current characteristics of electrical protective devices.

8) *PTW Equipment Evaluation* module compares protective device ratings with short-circuit calculations. The program also checks for missing input data and compares continuous ratings to calculated design and operating conditions. Equipment that fails the evaluation are reported in table form and color-coded by the one-line diagrams. As with all *PTW* study modules, Equipment Evaluation uses the same project database, integrating all balanced and unbalanced/single-phase study modules, and allowing you to examine existing projects without additional input requirements.

9) *PTW TMS*, Transient Motor Starting Analysis is a state-of-the-art time simulation program to analyze all aspects of motor starting problems accurately. TMS models up to 1500 motors dynamically throughout starting, stopping or reacting to load changes. In order to completely examine motor starting problems, TMS has the capability to dynamically represent motors which are already on line at the beginning of the simulation.

10) *PTW HI_WAVE*, Harmonic Analysis Program simulates resonance and harmonic distortion in industrial, commercial, and utility power systems.

11) *PTW Unbalanced Single Phase Studies* simulates systems with single-phase, two-phase and unbalanced three-phase load conditions. Phase and sequence currents can be displayed for different operating and load conditions including open-phase and simultaneous faults. Studies include demand load analysis, sizing, load flow/voltage drop and short-circuit. Reports also include three-phase and single-phase panel schedules. Modeling includes single-phase, two-phase, and three-phase lines, transformers, loads, and capacitors as well as single-phase mid-tap transformers.

12) *PTW I*SIM* is a program for transient stability analysis. It is designed to simulate system response during and after transient disturbances such as faults, load changes, switching, motor starting, loss of utility, loss of generation, loss of excitation, and blocked governor events. I*SIM is designed to study today's most challenging simulation problems in one convenient and easy-to-use program.

13) *PTW Reliability* calculates reliability indices and cost effects for alternative system designs. Calculations include alternative supplies, alternative network configurations, spare equipment, time to repair, and cost impact of lost production. Libraries for time-adjusted component failure rates and costs are provided to save time and simplify system modeling.

14) *PTW DC System Analysis* includes: Battery Sizing, DC Load Flow, DC Short-Circuit (ANSI) and DC Short-Circuit (IEC).

15) *PTW GroundMat* is a program for substation ground grid design and analysis. It is designed to help optimize grid design or reinforce existing grids of any shape. It uses a general-purpose finite element algorithm for potential analysis and graphical facilities to validate ground system efficiency.

16) *PTW CABLE 3D* quickly solves complex three-dimensional cable pulling tension and sidewall pressure calculations, allowing you to make rapid and accurate design decisions.

17) PTW Viewer is the perfect tool to provide an interactive electrical model to your customers. Rather than paging through long reports, your customers can display results on the interactive one-line, view time-current curves, print arc flash labels, and many other activities. With the Viewer there is no chance of inadvertent modifications to the power system model."

2.2 OBTAINING *SKM*

SKM can be obtained directly from its main office at *SKM* Systems Analysis, Inc. or from overseas distributors. *SKM* sells to most countries directly from its main office.

> *SKM* Systems Analysis, Inc.
> P.O. Box 3376
> Manhattan Beach, CA 90266
> 1-310-698-4700
> 1-800-232-6789
> Sales@*SKM*.com
> www.*SKM*.com

SKM is sold in modules. These are listed in Section 2.1. The purchaser first determines how many buses he will have in his circuit. Then he buys a combination package with at least that bus handling capability and buys whatever modules he would like to go with it. Alternatively, the purchaser could determine the number of buses he anticipates and then purchase a "Combo-Pack". The "Combo-Pack"'s contain the most useful program modules.

SKM offers a 30 day free *PTW* trial Demo. The Demo can be downloaded or received on a mailed CD. If you request the mailed Demo CD you will also received a printed tutorial, a CD with training videos, and other useful printed information.

All of this section's examples were done with the *SKM* 6.5 Demo.

2.3 COMPUTER SYSTEM REQUIREMENTS FOR *SKM* 6.5

Operating System
> *Microsoft Windows Vista,*
> *Microsoft Windows XP,*
> *Microsoft Windows 2003,*
> *or Microsoft Windows 2000*

PC Configuration Requirements
> CD-ROM Drive
> 4 GB of free hard disk space
> Printer or plotter recommended

Hardware Requirements
> *Intel Pentium III* 600 MHz Processor or better
> 512 MB of RAM (1 GB for 2000+ bus projects)

2.4 *SKM* DEMO LIMITATIONS

1) It is operational for 30 days. After that, contact *SKM* to ask for an extension.

2) A maximum of 15 buses and 35 components can be evaluated.

3) Circuits and data can be saved only to the *SKM* Demo program's database.

4) All modules are active except for *PTW GroundMat, PTW CABLE 3D,* and *ArcCalc.* Demos for these programs may be requested by contacting *SKM.*

5) Printing is disabled.

6) There are other limitations that would be of less interest to the beginner.

2.5 *SKM* TRAINING AND TUTORIALS

SKM offers a free printed tutorial titled *Power*Tools for Windows Tutorial*, nine free video tutorials on the *PTW* Tutorial CD, and for-a-fee classroom training. Information on these can be found on the *SKM* website, www.*SKM*.com.

The beginner should definitely read and do some of the exercises in the *Power*Tools for Windows Tutorial*. The *PTW Overview* video, from the *PTW Tutorial* CD, should be viewed. The other videos are also helpful.

SKM classroom training is done at their home office and in many locations around the U. S. and Canada For five or more students, they also offer on-site training. Live online training is also available.

2.6 SETUP AND INSTALLATION OF THE *SKM* DEMO

The Demo can be downloaded or received on a CD. Once the program file is received, just follow the prompts. No activation code is necessary. Partway through the installation there is a choice of IEC or ANSI systems. To configure *SKM* the same as it was for this book, select ANSI Engineering Symbols with English units.

2.7 EXAMPLE SYSTEM ANALYZED WITH *SKM*

The same example as was analyzed manually in Section 1.3 is analyzed here with *SKM* using its *PTW-DAPPER* program.

2.7.1 SETTING UP THE *SKM* DEMO FOR USE ON THE SYSTEM OF FIGURE 1-3-1

1) Start *SKM* by double left-clicking on the *SKM* "PTW32" icon.

2) The windows shown in Figure 2-7-1-1 appear.

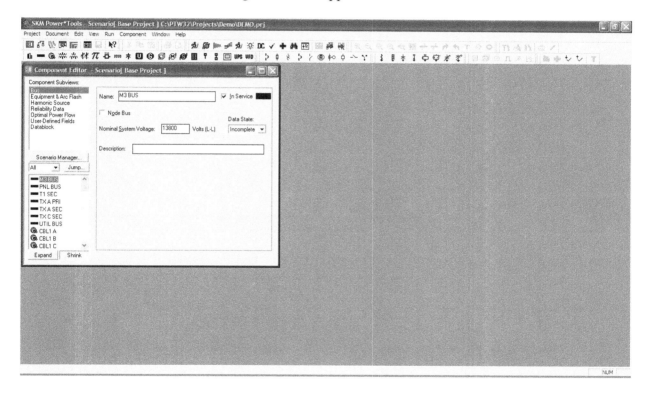

Figure 2-7-1-1 "Example Project (ANSI)" system.

3) On the screen's top left, left-click on "Project" and then left-click on "Close".

4) Again left-click on "Project" but this time then left-click on "New".

5) Type in the file name "exampleSKM" and save to *SKM*'s project database. *SKM*'s Demo version only saves project files to its own "Projects" directory.

6) Now *SKM* is ready to receive a one-line diagram and other project information into its "exampleSKM" project database. Each *SKM* project has its own database. The windows of Figure 2-7-1-2 appear.

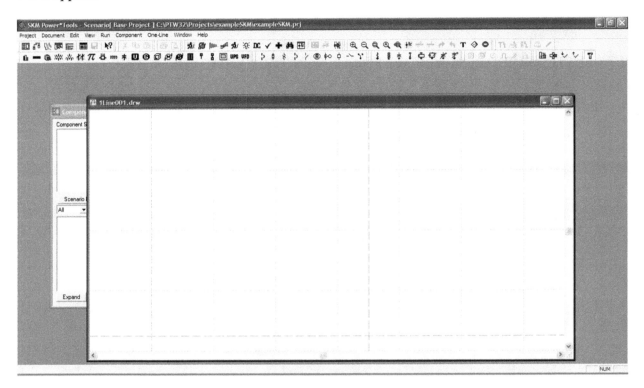

Figure 2-7-1-2 Work window ready for the entry of the Figure 1-3-1 system.

2.7.2 CREATING A ONE-LINE DIAGRAM

1) The one-line diagram of Figure 1-3-1 will be re-created below with *SKM*.

2) Left-click on the "New Utility" icon in the "Component" toolbar (the toolbar on the upper left side with transformer symbols on it), release the button (do not hold the left mouse button down, as would be done with some CAD programs), drag the "Utility" icon to a central location on the work window, and then left-click again to set it in place. After an icon is placed on the one-line diagram, then it can be moved by left-clicking on it, holding the left button down, and then releasing the left button when the icon is in the proper location. See Figure 2-7-2-1.

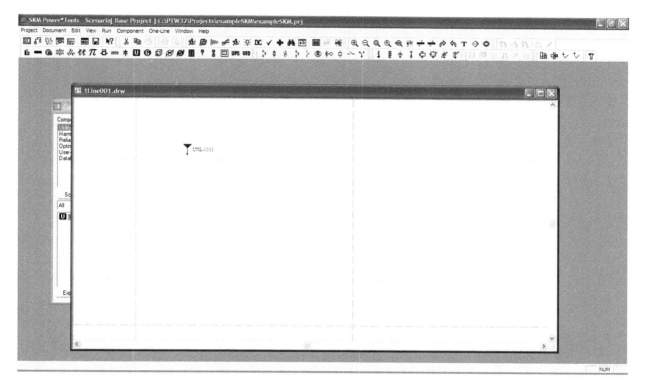

Figure 2-7-2-1 Work window with the "Utility" icon in place.

3) Save the one-line diagram by left-clicking on "Document" on the screen's upper left and then left-click on "Save". That will save it to its default name "1Line001.drw".

4) The "NEW BUS" icon is also in the "Component" toolbar. Left-click, drag, and set a "BUS" icon beneath the terminal of the work window's "Utility" icon.

5) With the cursor on the terminal of the "Utility" icon, left-click and hold down the mouse button. Drag a line straight down to the "BUS" icon. Release the left mouse button and the "Utility" icon will be connected to the "BUS" icon by a solid line.

6) Left-click and drag the other icons to the work window and then connect them. When needed, the "BUS" icon can be stretched by left-clicking on an end and dragging it. The work window can stretched out to create a larger viewing area. The drawing size can be zoomed in or out using "Zoom" under the "View" menu or by turning the mouse's wheel on a mouse equipped with a wheel. The resulting one-line diagram is shown in Figure 2-7-2-2.

Note:
1) *SKM* will not allow two impedance components in series without a "BUS" between them. Cables and transformers are examples of impedance components.
2) *SKM* will not allow "BUS"es to be connected by zero impedance devices. Fuses and circuit breakers are examples of zero impedance components.

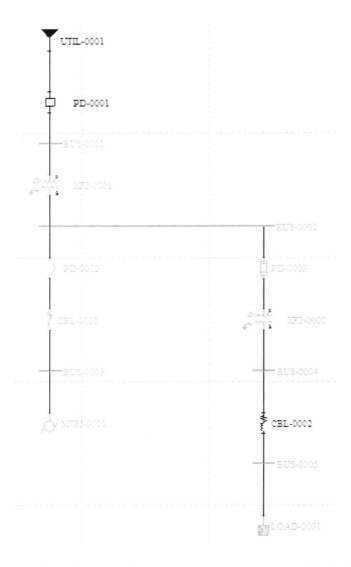

Figure 2-7-2-2 *SKM* one-line diagram of the Figure 1-3-1 example with default transformer types. No system data has been entered.

2.7.3 LOAD FLOW ANALYSIS

2.7.3.1 Circuit Data Entry

SKM has many data windows for each component. With these each component can be described with a great amount of data. However, most of it is not needed by *SKM* for load flow and balanced short-circuit analyses. Here, only the required data will be entered.

Underlined values are entered into *SKM*.

2.7.3.1.1 Bus data for "BUS-0001", "BUS-0002", "BUS-0003", "BUS-0004", and "BUS-0005"

   1) Double left-click on the one-line diagram's "BUS-0001" icon. Do not click on the text "BUS-0001" or else a "Rename Component" window will appear. Manual analysis Section 1.3.1.2.1 shows that the value of the bus's nominal line-to-line voltage is 3800 V rms. Enter 3800 into the "Bus Component Subview" window. Once a value has been entered, simply exit out of the window. *SKM* automatically saves the value. The "BUS-0001" bus data window is shown in Figure 2-7-3-1-1-1.

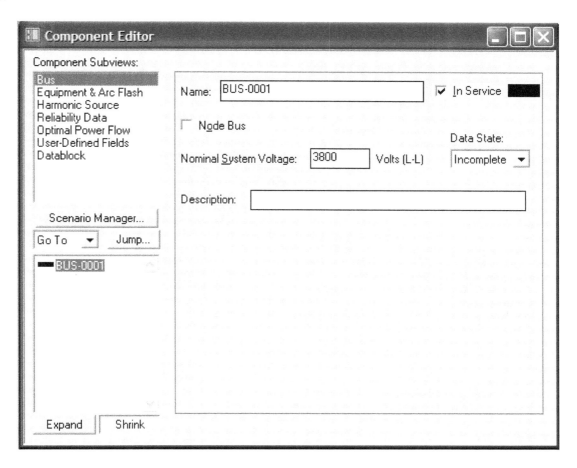

Figure 2-7-3-1-1-1 "Component Editor" "BUS-0001" window.

   2) In the same way manual analysis nominal voltage values from Sections 1.3.1.1.1 and 1.3.1.1.2 are entered into the windows for "BUS-0002" and "BUS-0004". The voltages are 480 and 120 V rms line-to-line respectively. Voltages do not have to be entered for "BUS-0003" and "BUS-0005". *SKM* will assign the nominal voltages of "BUS-0002"and "BUS-0004" to them.

2.7.3.1.2 Data for the cable from circuit breaker C2 to motor M, "CBL-0001"

1) From the manual analysis information in Section 1.3.1.3.3:

Length = 200 ft.
Positive Sequence Resistive Impedance = .000078 Ω/ft. = 0.078 Ω/1000 ft.
Positive Sequence Inductive Impedance = .000030 Ω/ft. = 0.030 Ω/1000 ft.

2) Double left-click on the "CBL-0001" icon to open the cable's "Component Editor". Enter the 200 ft. length on the "Cable Component Subview" window. See Figure 2-7-3-1-2-1.

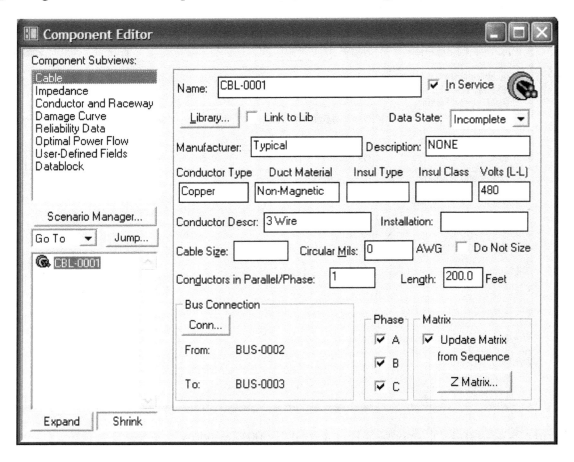

Figure 2-7-3-1-2-1 "Component Editor" "Cable Component Subview" "CBL-0001" window.

3) From the "Cable Component Subview" left-click on "Impedance". Then enter the positive sequence impedance per 1000 ft. Negative and zero sequence impedances are not needed for load flow and balanced short-circuit calculations. However, by default *SKM* automatically makes the negative and zero sequence impedances the same as the positive sequence impedances. The cable impedance data window is shown in Figure 2-7-3-1-2-2.

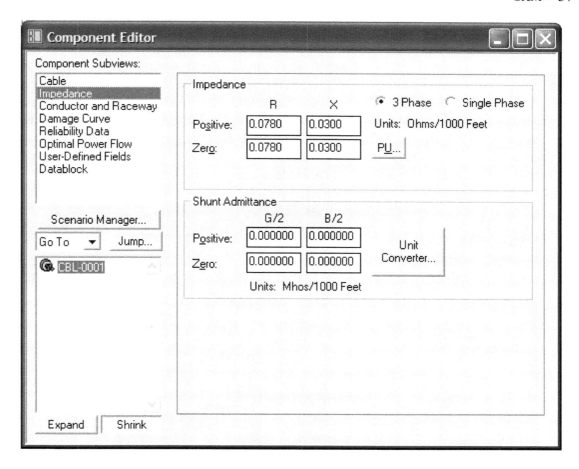

Figure 2-7-3-1-2-2 "Component Editor" "Impedance Subview" "CBL-0001" window.

2.7.3.1.3 Data for the cable from transformer T2 to static load R, "CBL-0002"

1) From the manual analysis information in Section 1.3.1.3.4:

Length = <u>200</u> ft.
Positive Sequence Resistive Impedance = .000098 Ω/ft. = <u>0.098</u> Ω/1000 ft.
Positive Sequence Inductive Impedance = .000030 Ω/ft. = <u>0.030</u> Ω/1000 ft.

2) The cable data is entered here as it was in 2.7.3.1.2

2.7.3.1.4 Transformer T1 data

1) Transformer T1 data is taken from the manual analysis information in Section 1.3.1.1.1. The transformer's input and output voltages, 3800 and 480, were already automatically entered when the bus voltages were entered. The winding connections can be left at their default connections, Δ-Y(ground) for load flow and balanced short-circuit analysis. The transformer winding connections do not matter for load flow and balanced short-circuit analysis, but will need to be corrected for unbalanced short-circuit analysis in Section 2.7.5. The transformer's nominal kVA rating of <u>600</u> kVA is entered. The 2-winding transformer data window is shown in Figure 2-7-3-1-4-1.

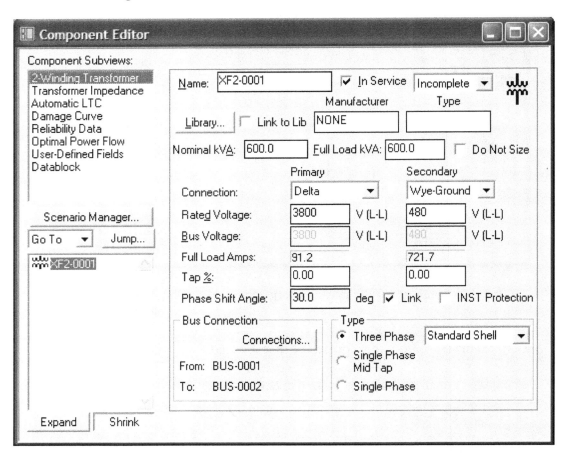

Figure 2-7-3-1-4-1 Transformer T1 "Component Editor" "2-Winding Transformer Component Subview" window with default (not yet corrected for this example) winding connections.

2) Select the "Transformer Impedance Component Subview". Then enter % positive sequence impedances. The base for the % values is the transformer's nominal rating. The impedances can be taken directly from the manual analysis in Section 1.3.1.1.1. They are:

Positive Sequence Resistive Impedance in Percent on Transformer Base = <u>1.73</u>%
Positive Sequence Inductive Impedance in Percent on Transformer Base = <u>1.52</u>%

By default *SKM* automatically inputs the same values for transformer negative and zero sequence impedances as for positive sequence impedances. These values do not affect load flow and balanced short-circuit analyses.

The transformer impedance data window is shown in Figure 2-7-3-1-4-2.

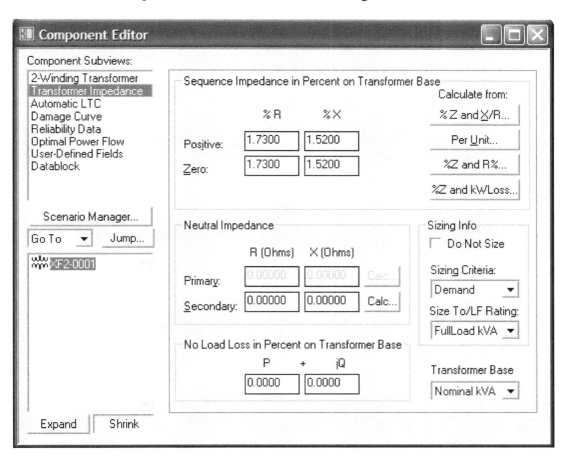

Figure 2-7-3-1-4-2 Transformer T1 "Component Editor" "Transformer Impedance Component Subview" window.

2.7.3.1.5 Transformer T2 data

1) Transformer T2 data is taken from the manual analysis information in Section 1.3.1.1.2. The transformer's input and output voltages, 480 and 120, were already automatically entered when the bus voltages were entered. The transformer's nominal kVA rating of 90 kVA is now entered. Data is entered as it was in Section 2.7.3.1.4.

Positive Sequence Resistive Impedance in Percent on Transformer Base = 2.00%
Positive Sequence Inductive Impedance in Percent on Transformer Base = 1.80%

2) The transformer data is entered here as it was in Section 2.7.3.1.4.

2.7.3.1.6 Utility data

1) Impedances are taken from the manual analysis information in Section 1.3.1.2.2.

Positive Sequence Resistive Impedance = .2 Ω
Positive Sequence Inductive Impedance = .3 Ω

2) The impedances are converted to per-unit values with a 100 MVA base. The values that *SKM* will accept are calculated below and underlined:

Positive Sequence Per-Unit Resistive Impedance Contribution =
(Positive Sequence Resistive Impedance) (Power base/VLL^2) =
$.2 \times (100 \times 10^6/3800^2) = 1.385$

Positive Sequence Per-Unit Inductive Impedance Contribution =
(Positive Sequence Inductive Impedance) (Power base/VLL^2) =
$.3 \times (100 \times 10^6/3800^2) = 2.078$

3) The manual analysis 3800 voltage of Section 1.3.1.2.1 was previously automatically entered when 3800 was entered on "BUS-0001".

4) For load flow analysis and balanced short-circuit analysis, values for the negative and zero sequence impedances and connection configuration are not needed and can be left at default values.

5) The utility data window is shown in Figure 2-7-3-1-6-1.

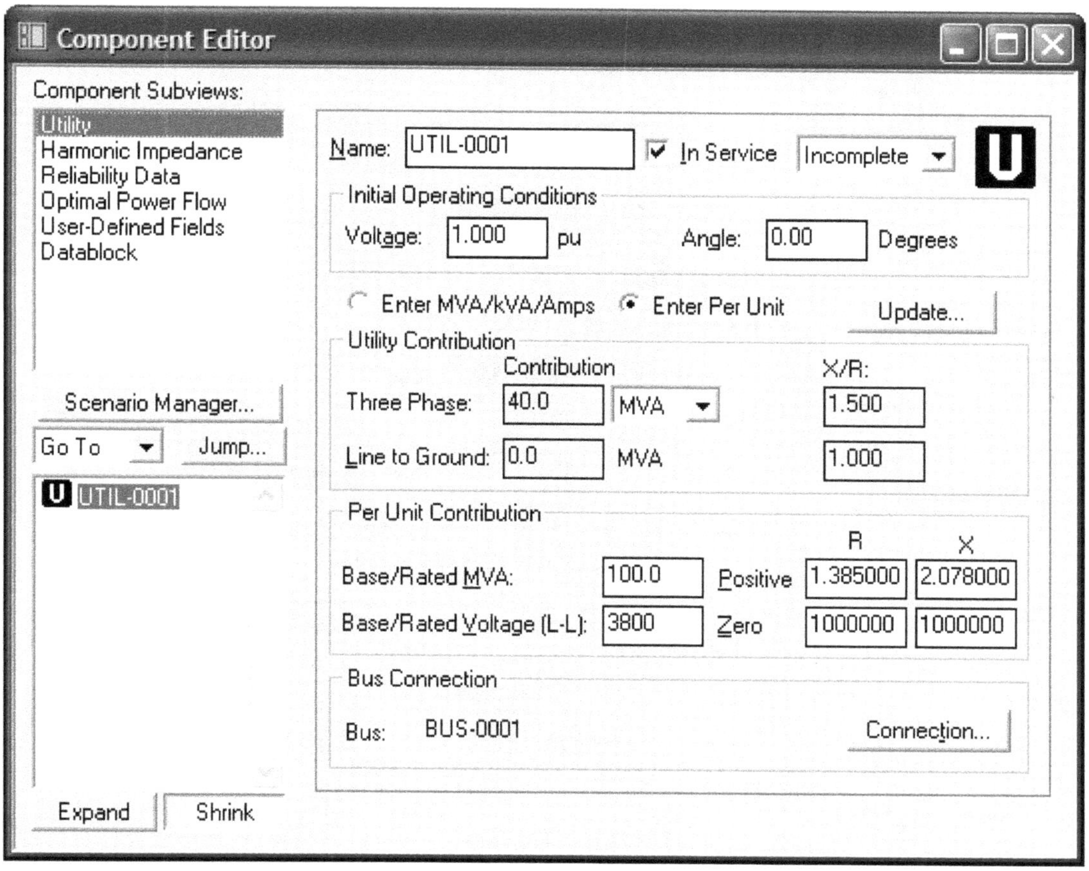

Figure 2-7-3-1-6-1 Utility "Component Editor" "Utility Component Subview" window.

2.7.3.1.7 Motor M data

1) Motor M horsepower is taken from the manual analysis information in Section 1.3.1.4.

Rated Size = <u>250</u> hp
Power Factor = <u>.918</u>
Efficiency = <u>.9286</u>

2) The 480 volt motor voltage was already automatically entered when the "BUS-0002" voltage was specified.

3) The induction motor M data window is shown in Figure 2-7-3-1-7-1.

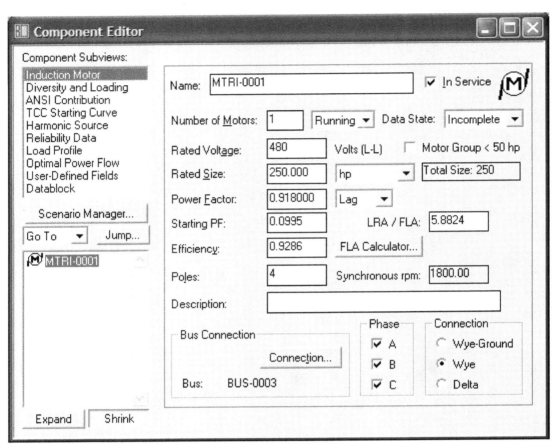

Figure 2-7-3-1-7-1   Motor M "Component Editor" "Induction Motor Component Subview" window.

2.7.3.1.8 General load R data

1) General load R data is taken from the manual analysis information in Section 1.3.1.5. The underlined data can be directly input to the *SKM* program.

> Rated Size = <u>30.0</u> kVA
> Power Factor = <u>Unity</u>

2) For load flow analysis and balanced short-circuit analysis, values for the negative and zero sequence impedances and connection configuration are not needed and can be left at default values.

3) The general load R data window is shown in Figure 2-7-3-1-8-1.

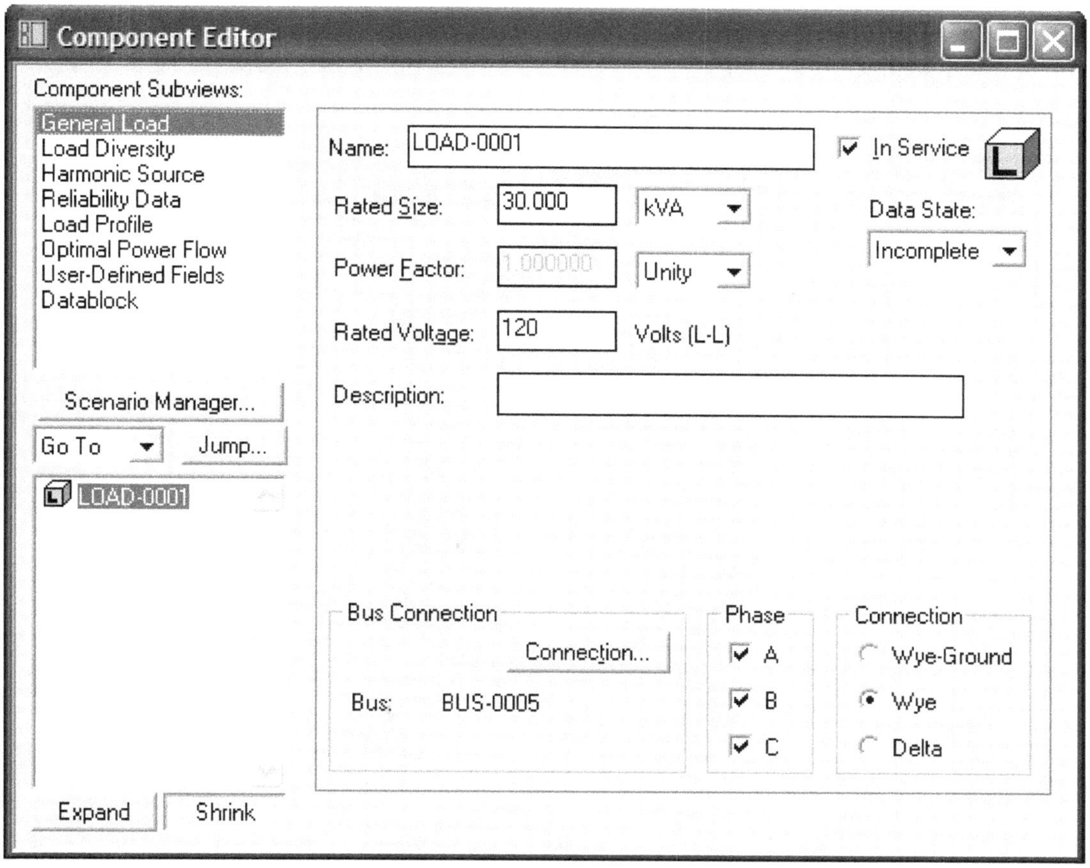

Figure 2-7-3-1-8-1 "General Load" R "Component Editor" "General Load Component Subview" window.

2.7.3.1.9 Circuit breaker C1 and C2 and fuse F1 data

No data input is necessary.

2.7.3.2 Load flow Analysis

1) Left-click on the icon in the "Tool" toolbar that looks like a person sitting at computer in front of a cyan colored background. This is titled "Balanced System Studies". The window in Figure 2-7-3-2-1 appears.

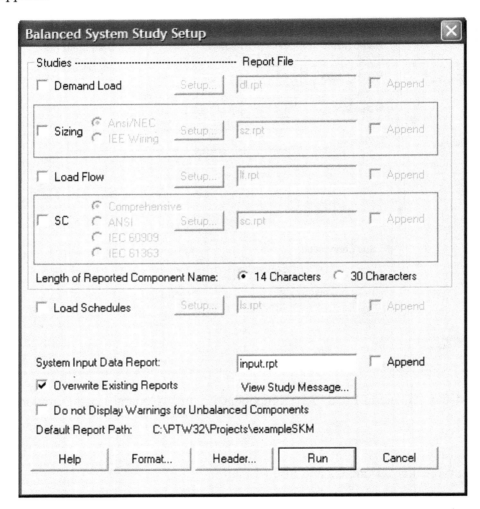

Figure 2-7-3-2-1 "Balanced System Study Setup" window.

2) Left-click on "Load Flow" and "Run".

3) A "Study Messages" window appears. See Figure 2-7-3-2-2.

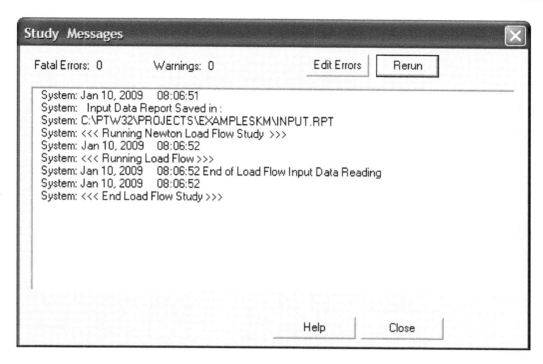

Figure 2-7-3-2-2 "Study Messages" window.

4) Left-click "Close" on the "Study Messages" window.

5) Left-click on "Run" in the upper menu and then left-click on "Datablock Format..." The window in Figure 2-7-3-2-3 appears.

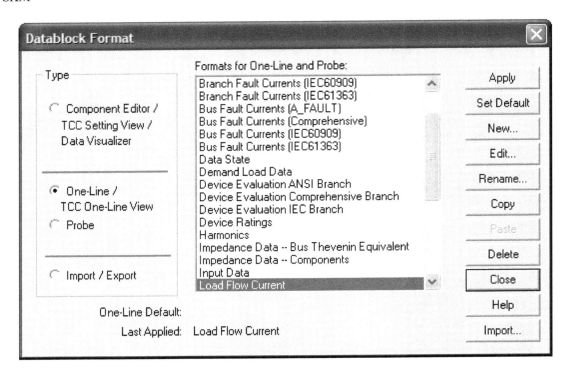

Figure 2-7-3-2-3 "Datablock Format" window.

6) Left-click on "One-Line/TCC One-Line View", "Load Flow Current", and then on "Apply". Move the "Datablock Format" out of the way to see the one-line diagram with load flow currents. See Figure 2-7-3-2-4.

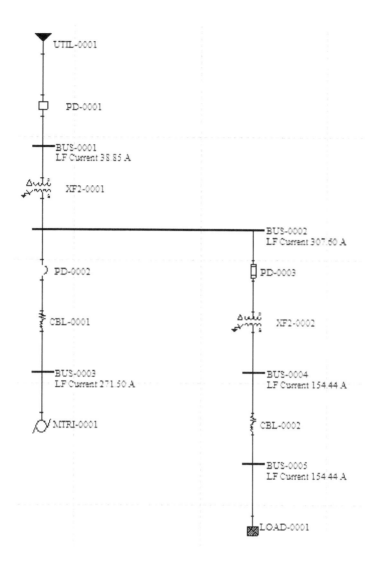

Figure 2-7-3-2-4 One-line diagram with load flow currents.

8) Select "Load Flow Voltage" in "Datablock Format" and follow the same procedures as in 6). This will produce Figure 2-7-3-3-5.

Note:

There are many quantities available in the "Datablock Format" window and even more quantities can be imported to the "Datablock Format" by left-clicking on "New".

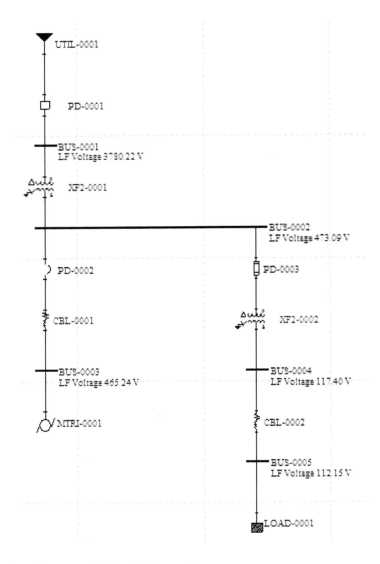

Figure 2-7-3-2-5 One-line diagram with load flow voltages.

2.7.4 BALANCED SHORT-CIRCUIT ANALYSIS

1) As with the load flow analysis of Section 2.7.3, left-click on the icon in the "Tool" toolbar that looks like a person sitting at computer in front of a cyan colored background. This is titled "Balanced System Studies".

2) The "Balanced System Study Setup" window appears. On this, left-click on "SC" <Short-Circuit> and "Comprehensive". See Figure 2-7-4-1.

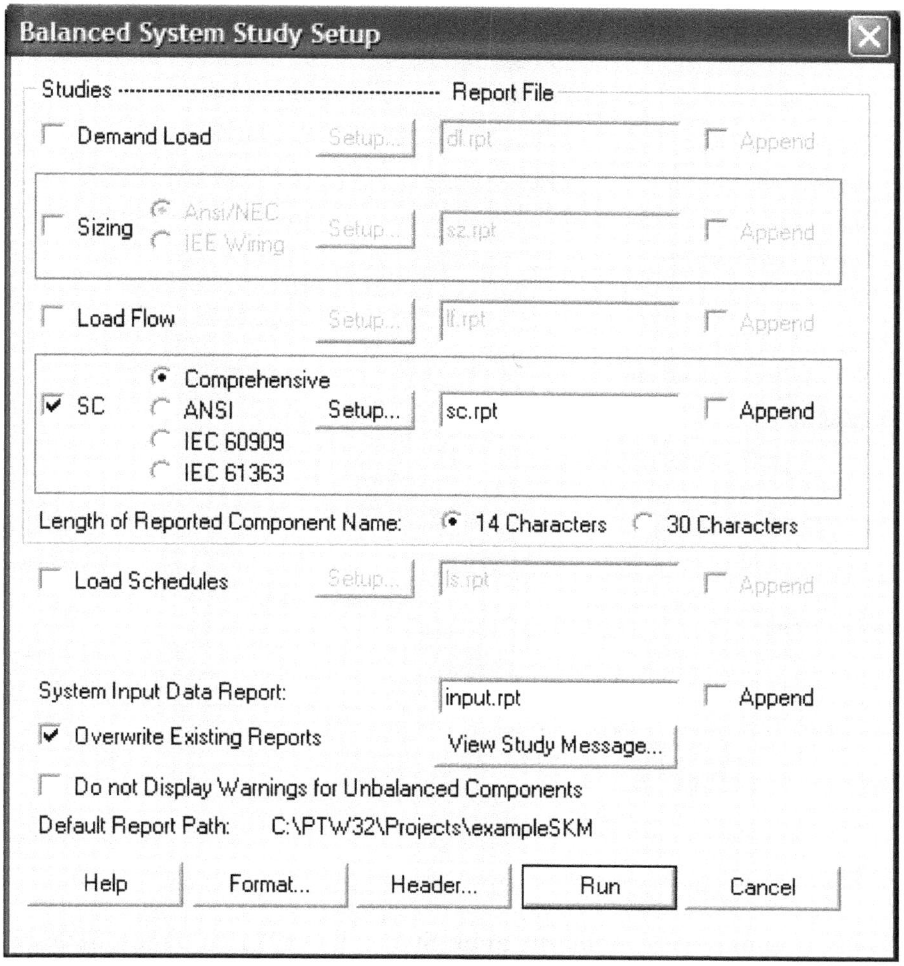

Figure 2-7-4-1 "Balanced System Study Setup".

3) In the "SC" box left-click on "Setup…". The "Comprehensive Short Circuit Study" window appears. In this left-click on "Three Phase Fault", "All Buses", and "OK" as shown in Figure 2-7-4-2.

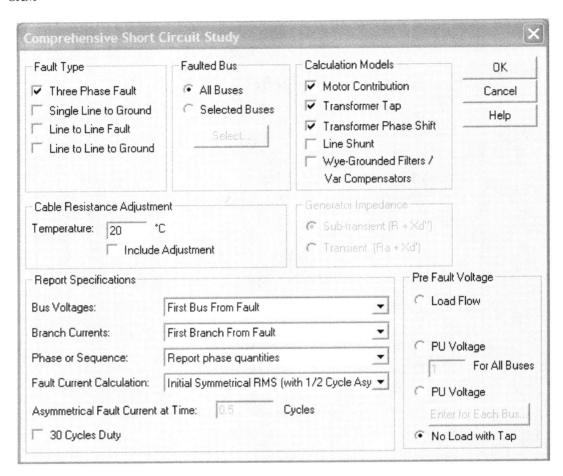

Figure 2-7-4-2 "Comprehensive Short Circuit Study" for balanced three-phase short circuits.

4) On the "Balanced System Study Setup" left-click on "Run". Move the "Study Messages" window out of the way or delete it. The one-line diagram now shows balanced three-phase short circuits as they would occur at each bus. Each current was determined for one short circuit at a time. The current values do not represent simultaneously occurring short circuits. See Figure 2-7-4-3.

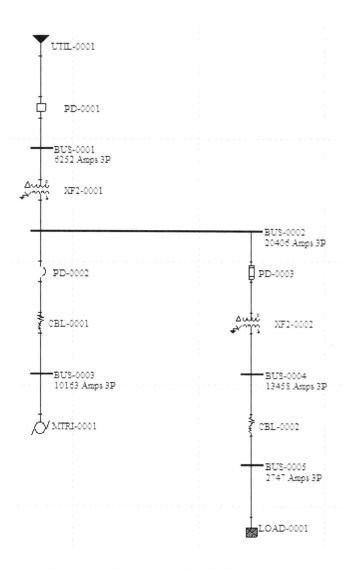

Figure 2-7-4-3 "Bus Fault Currents (Comprehensive)". This gives the balanced three-phase short-circuit currents that would occur at each bus. The short-circuit current values represent one short circuit at a time.

2.7.5 UNBALANCED SHORT-CIRCUIT ANALYSIS

Unbalanced short-circuit analysis requires more information than load flow and balanced short-circuit analysis require. The voltage supply and the impedances used in the load flow and balanced short-circuit analyses are *symmetrical components* positive sequence impedances. For unbalanced short-circuit analysis *symmetrical components* negative and zero sequence impedances are also needed. Furthermore, unlike balanced analyses, the *SKM* one-line diagram needs correct transformer connections (i.e. Y-Y, Y-Y(ground), Y-Δ, etc.).

2.7.5.1 Data for the Cable from Circuit Breaker C2 to Motor M, "CBL-0001"

SKM assumes the negative and zero sequence impedances are the same as the positive. The zero sequence impedances could be adjusted if necessary. However, in this example circuit the zero sequence impedances are the same as the positive, so nothing needs to be changed.

2.7.5.2 Data for the Cable from Transformer T2 to Static Load R, "CBL-0002"

As with the cable in Section 2.7.5.1, nothing needs to be changed.

2.7.5.3 Transformer T1 Data

1) Transformer T1 data is taken from the manual analysis information in Section 1.3.1.1.1.

2) As with the cables, *SKM* assumes that the negative sequence and zero sequence impedances are the same as the positive. Again the zero sequence impedance could be changed if desired, but that is not necessary here. It has the same values as the positive.

3) To correct the transformer winding connections, return to the one-line diagram, double left-click on the transformer, and enter the correct winding connections on the "Component Editor" "2-winding Transformer Component Subview". This is shown for transformer T1 in Figure 2-7-5-3-1.

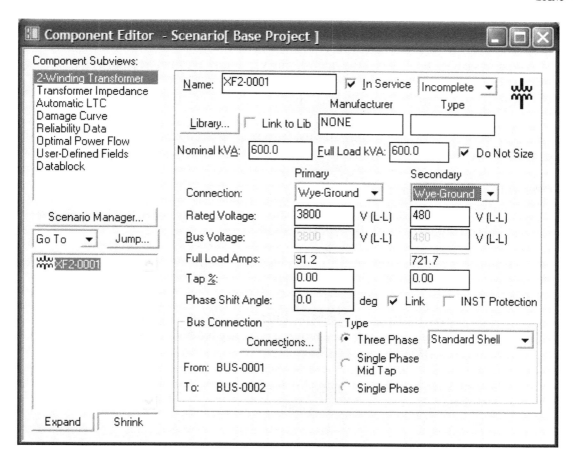

Figure 2-7-5-3-1 Transformer T1 "Component Editor" "2-Winding Transformer Component Subview" with correct winding connections.

2.7.5.4 Transformer T2 Data

1) As with T1, the zero sequence impedance does not have to be changed.

2) Using the "Component Editor" "2-Winding Transformer Component Subview" change the winding connections to Y(ground)-Y.

2.7.5.5 Utility Data

1) Impedances are taken from the manual analysis information in Section 1.3.1.2.2.

2) *SKM* assumes that the negative sequence impedances are the same as the positive.

3) *SKM* assumes very large values for the zero sequence impedances. These must be changed to agree with supplied values.

The values from Section 1.3.1.2.2 are:
Zero Sequence Resistive Impedance = .1 Ω
Zero Sequence Inductive Impedance = .15 Ω

4) The impedances are converted to per-unit values with a 100 MVA base. The values that *SKM* will accept are calculated below and underlined:

Zero Sequence Per-Unit Resistive Impedance Contribution =
(Zero Sequence Resistive Impedance) (Power base/VLL^2) =
.1x(100 x 10^6/3800^2) = .6925

Zero Sequence Per-Unit Inductive Impedance Contribution =
(Zero Sequence Inductive Impedance) (Power base/VLL^2) =
.15x(100 x 10^6/3800^2) = 1.039

4) The utility data window is shown in Figure 2-7-5-5-1.

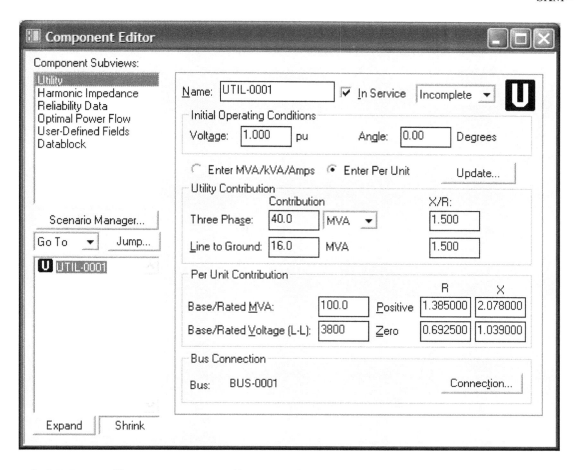

Figure 2-7-5-5-1 Utility "Component Editor" "Utility Component Subview" window.

2.7.5.6 Motor M Data

SKM selected typical positive, negative, and zero sequence impedances using the already input rated motor horsepower, power factor, efficiency, connection, and voltage. It is not necessary to enter more data here.

2.7.5.7 General Load R Data

No data input is necessary.

2.7.3.8 Circuit Breaker C1 and C2 and Fuse F1 Data

No data input is necessary.

2.7.5.9 Unbalanced Short-Circuit Analysis of a Line-to-Ground Short Circuit

1) As with the load flow analysis of Section 2.7.3, left-click on the icon in the "Tool" toolbar that looks like a person sitting at computer in front of a cyan colored background. This is titled "Balanced System Studies".

2) The "Balanced System Study Setup" window appears. On this, left-click on "SC" <Short Circuit> and "Comprehensive". See Figure 2-7-5-9-1.

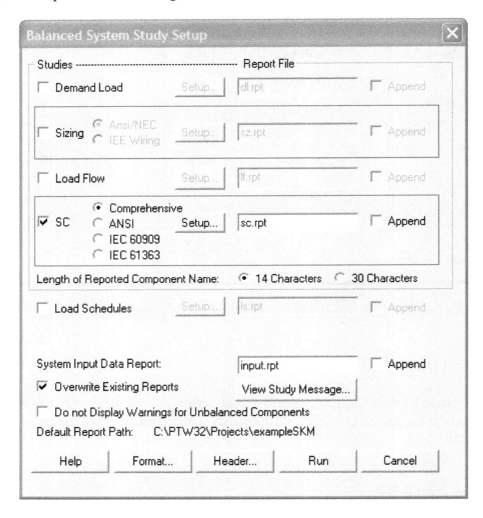

Figure 2-7-5-9-1 "Balanced System Study Setup" window.

3) On the "Balanced System Study Setup" window left-click on "Setup…" in the "SC" box. This will cause the "Comprehensive Short Circuit Study" window to appear.

SKM--- 57

4) In the "Comprehensive Short Circuit Study" window select "Single Line to Ground" and "All Buses". Then left-click on "OK". See Figure 2-7-5-9-2.

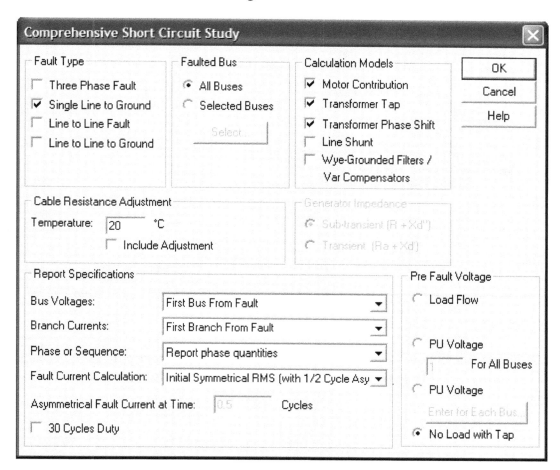

Figure 2-7-5-9-2 "Comprehensive Short Circuit Study" window.

5) The "Balanced System Study Setup" window reappears. On it, left-click on "Run".

6) The "Study Messages" window of Figure 2-7-5-9-3 appears.

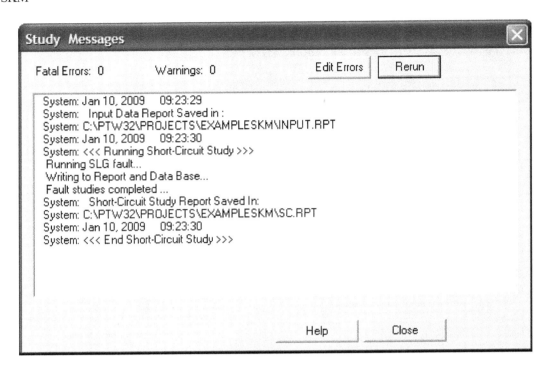

Figure 2-7-5-9-3 "Study Messages" window.

7) Delete the "Study Messages" window. This reveals the one-line diagram beneath it with single-line-to-ground, "SLG", short-circuit currents at buses that have a ground path. This can be seen in Figure 2-7-5-9-4.

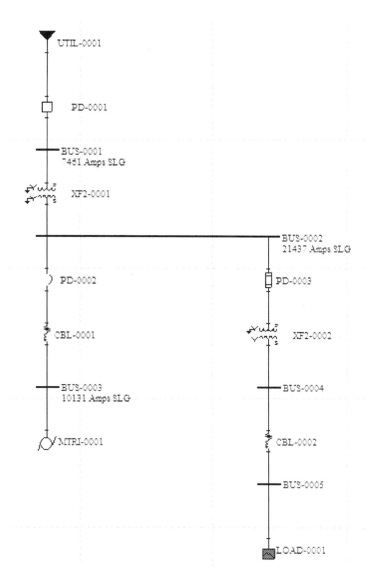

Figure 2-7-5-9-4 One-line diagram with single-line-to-ground, "SLG", short-circuit currents.

2.8 *SKM* REFERENCES

1) *SKM* Systems Analysis, Inc., *Power*Tools for Windows Tutorial*, 2008. It is 360 pages long and 8.5" x 11". A person learning *SKM* should get a copy. It is currently available for free on http://www.SKM.com/PTW V6.5_Tutorial.pdf

2) *SKM* Systems Analysis, Inc., *PTW Tutorial* video. This is a DVD disk meant for running on *Windows*. It is free and comes with the Demo disk. A person learning *SKM* should get a copy. It is available for free at http://www.skm.com/af.shtml

3.0 *ETAP*

ETAP is an acronym for **E**lectrical **T**ransient **A**nalysis **P**rogram. It was first made available in 1986.

3.1 FEATURES, AS STATED BY *ETAP*

The 2008 *ETAP* Product Overview sales brochure states:

"*ETAP* is the most comprehensive analysis platform for the design, simulation, and operation of generation, distribution, and industrial power systems. *ETAP* is developed under an established quality assurance program and is used as a high impact program worldwide. As a fully integrated enterprise solution, *ETAP* extends to a Real-Time Intelligent Power Management System to monitor, control, automate, simulate, and optimize the operation of your system."

ETAP sales staff says advantages of their program are:

"1) *ETAP* has a three dimensional database which allows for easy 'what-if' scenarios. *ETAP* can store unlimited 'what-if' scenarios in one integrated common database. Our competition can only accomplish this with multiple database copies.

2) *ETAP* allows for composite networks and unlimited nested one-line diagrams.

3) *ETAP* automatically tells you if your equipment is correctly sized via graphical reports.

4) 88% of nuclear power facilities are standardized on *ETAP*. *ETAP* is the only NRC certified power system software provider.

5) *ETAP* has 45,000 licenses worldwide, making it the largest power system software supplier in the world.

6) Customers accustomed to competitor's software say that *ETAP*'s graphics are leaps and bounds ahead of the competition and that *ETAP* back annotation is more presentable.

7) You can export *ETAP* files to *AutoCAD* DXF files. Other programs, at best, only allow export of image files.

8) You can import existing Ground Grid Systems created in *AutoCAD* to *ETAP*.

9) Easily interfaces with *ETAP* Real-Time Systems such as ILS, PSMS, & EMS."

The 2008 *ETAP* Product Overview sales brochures list the following *ETAP* program modules:

"1) *Base Package:*

ETAP Base Package is a set of core tools, embedded analysis modules, and engineering libraries that allow you to create, configure, customize, and manage your system model. Core tools allow you to quickly and easily build 3-phase and 1-phase AC and DC network one-line diagrams with unlimited buses and elements including detailed instrumentation and grounding components. Base Package includes an intelligent one-line diagram, element editors, configuration manager, report manager, project and study wizards, multi-dimensional database, theme manager, data exchange, and user access management. Embedded analysis modules such as Cable Ampacity, Cable Sizing, and Transmission Line Constants provide integrated as well as stand-alone capabilities to design, analyze, and size equipment. Engineering libraries provide complete verified and validated data based on equipment manufacturer's published data.

2) *Simulation & Analysis:*

Short-Circuit ANSI/IEEE/IEC: Save hours of tedious hand calculations and take the guesswork out of short circuit analysis by automating the process with multiple calculation and result analysis tools within *ETAP*.

The Short-Circuit module makes it easy to go from selecting elements from the comprehensive libraries of short circuit current ratings to performing dozens of different types of short circuit analysis with the purpose of finding the worst-case device duty. Built-in intelligence allows it to automatically apply all ANSI / IEEE or IEC factors and ratios required for high and low voltage device duty evaluation. Determine short-circuit currents and automatically compare these values against manufacturer short circuit current ratings. Overstressed device alarms are automatically displayed on the one-line diagram and reports.

OTI closely monitors industry standards and OEM data to ensure *ETAP* is always up-to-date. The Short-Circuit module seamlessly integrates with device coordination and performs arc flash hazard calculations.

Arc Flash: *ETAP* Arc Flash Analysis software brings you new and enhanced capabilities which allow for faster and easier performance of arc flash hazard analysis. Identify and analyze high risk arc flash areas in your electrical system with greater flexibility by simulating various incident energy mitigation methods.

Arc Flash is a completely integrated module that solves multiple scenarios to determine worst-case incident energy levels. It also produces professional reports and high quality arc flash labels at the press of a button.

Load Flow: With *ETAP*'s advanced load flow analysis software module, you can create and validate your system model with ease and obtain accurate and reliable results. Built-in features like automatic device evaluation, summary alarms / warnings, result analyzer, and intelligent graphics make it the most efficient power flow analysis program available today.

ETAP load flow analysis software calculates bus voltages, branch power factors, currents, and power flows throughout the electrical system. *ETAP* allows for swing, voltage regulated, and unregulated power sources with multiple power grids and generator connections. It is capable of performing analysis on both radial and loop systems. *ETAP* allows you to select from several different methods in order to achieve the best calculation efficiency and accuracy.

Motor Acceleration: The Motor Acceleration module enables engineers to thoroughly evaluate the impact of load changes to electric power systems. Motor Acceleration is fully capable of starting one motor or transitioning an entire power system to another state. Sequence-start a series of machines using static or dynamic models, operate Motor Operated Valves (MOVs), and simulate the switching actions of Load Tap Changers. Advanced plotting and time varying graphical display enable engineers to quickly evaluate results and make decisions.

Panel & Single Phase Systems: Developed for electrical designers and engineers, the Panel Systems module combines a graphical user interface and the intelligence of *ETAP* to easily design and analyze low voltage distribution systems.

Coupled with exclusive features and advanced capabilities, Panel Systems is a quality tool you would only expect from the leader in power system analysis software.

3) *Protective Devices:*

ETAP Star device protection and coordination program provides an intuitive and logical approach to Time Current Characteristic (TCC) analysis with features and capabilities such as graphical user interface, true-to-the-form protective devices modeling, extensive Verified & Validated (V&V) device library, embedded analysis modules, and integrated, rule-based design.

Using intelligent one-line diagrams, comprehensive device libraries, and a three-dimensional database, *ETAP* Star offers insight into troubleshooting false trips, relay mis-operation, and mis-coordination.

4) Relay Testing and Simulation:

The *ETAP* Advanced Relay Testing and Transient Simulator (ARTTS) is a new concept that utilizes hardware and software technologies for testing, calibrating, and simulating relays.

ETAP ARTTS combines the short circuit and protection device coordination capabilities of *ETAP* with the relay testing hardware. It provides actual steady-state and transient responses of relays for comparisons with the manufacturers published data. ARTTS is designed to improve system protection, coordination, and reliability, hence decreasing operational and maintenance costs.

5) *Cable Systems:*

Underground Thermal Analysis: Underground Thermal Analysis helps engineers to design cable systems to operate to their maximum potential while providing secure and reliable operation.

The advanced graphical interface allows for design of cable raceway systems to meet existing and future needs by using precise calculations to determine required cable sizes, their physical capabilities, and maximum derated ampacity.

In addition, transient temperature analysis computes temperature profiles for cable currents, reducing the risk of damage to cable systems under emergency conditions.

Cable Pulling: Accurate prediction of cable pulling forces is essential for the proper design of cable systems. This knowledge makes it possible to avoid under-estimated and/or over conservative design practices to achieve substantial capital savings during construction. The Cable Pulling module accounts for multiple cables of different sizes and allows complex 3-D pulling path geometry. A point-by-point calculation method is performed at every conduit bend and pull point. Both the forward and reverse pulling tensions are calculated for determining the preferred direction of pull.

Cable Ampacity: *ETAP* calculates cable ampacity based on NEC and ICEA P.54-440 methods for U/G duct banks, U/G direct buried, A/G cable trays, A/G conduits, and air drops. The process is systematic and simple. For example, for A/G trays, simply enter the tray height, width, and percent fill, *ETAP* calculates the derated ampacity based on user specified ambient and conductor operating temperatures. For duct banks, specify the number of rows, columns, ambient temperature, and soil thermals resistivity, *ETAP* calculates the derated ampacity based on the hottest location not exceeding the maximum operating temperature.

Cable Sizing: ETAP provides optimal and alternative cable sizes based on voltage drop and load current requirements. Load current can be based on the full-load amp of any element on the one-line diagram or as a user-specified value. You can size cables (motor feeders, transformer cables, etc.) instantly based on the cable derated ampacity for any type of installation (direct banks, trays, conduit in air, etc.).

6) *Power Quality:*

With *ETAP*'s Harmonic Analysis module, you can identify harmonic problems, reduce nuisance trips, design and test filters, and report distortion limit violations. Comprehensive load flow and frequency scan calculations are performed using detailed harmonic models and non-integer harmonic filters. Results are shown graphically, including harmonic order, harmonic spectrum plots, and harmonic waveform plots, as well as Crystal Reports.

7) *Dynamics & Transients:*

Transient Stability: The Transient Stability module enables engineers to accurately model power system dynamics and simulate system disturbances and events. Typical transient stability studies include identifying critical clearing time, motor dynamic acceleration, load shedding schedule, fast bus transfer timing, and generator start-up. You can split a system or combine multiple subsystems, simulate automatic relay actions and associated circuit breaker operations, accelerate or re-accelerate motors. Combined with enhanced plotting and graphical results, engineers can truly use this module to master power system stability studies.

Generator Start-Up: Using full frequency-dependent machine and network models, the Generator Start-Up module analyzes cold-state starting of generators under normal and emergency conditions. The entire generator start-up process is modeled, including automatic control relay simulation and the dynamic behavior of exciters/AVRs, governors, turbines, and Power System Stabilizers (PSS). You can simulate the starting of generators, connection of generators to the network before reaching synchronizing speed, acceleration of motors, action of MOVs, and operation circuit breakers.

Wind Turbine Generator: The Wind Turbine Generator (WTG) module allows you to design and monitor wind farms via a highly flexible graphic interface optimized for both steady-state and dynamic simulation. The WTG module is fully integrated with all *ETAP* calculation modules such as Load Flow, Short Circuit, Transient Stability, Harmonic Analysis, Protective Device Coordination, and *ETAP* Real-Time. User-defined actions may be added to simulate disturbances like wind variation and relay operations. It also predicts the dynamic response of each individual wind turbine generator. Analysis results may be utilized to analyze alternative turbine placement, tuning of control parameters, selection and placement of protective devices, and sizing associated equipment.

User-Defined Dynamic Model: User-Defined Dynamic Models (UDM) can be used to model or customize complex machine control systems. This module allows you to build control block diagrams needed to simulate the behavior of machines in Transient Stability and Generator Start-Up simulations. UDM provides independent self-testing via load rejection, load acceptance, and terminal bus faults for validation of models and their dynamic behavior.

Motor Parameter Estimation: The *ETAP* Parameter Estimation program calculates equivalent circuit model parameters for machines at starting conditions. The calculation is based on advanced mathematical estimation and curve fitting techniques, which require only the machine performance characteristic data.

The estimated model together with its parameters can be used to represent the machine dynamics during motor starting and transient stability studies. Machine characteristic curves based on the estimated model are automatically updated into the corresponding motor editor. Additional key machine characteristic and nameplate data are automatically calculated based upon the estimated model.

8) *Ground Grid Systems:*
   Finite Element Method
   IEEE 80 & 665 Methods

9) *Distribution Systems:*
   Unbalanced Load Flow
   Optimal Power Flow
   Optimal Capacitor Placement
   Reliability Assessment

10) *Geographic Information System:*
   GIS Map

11) *Transmission Line:*
   Line Constants
   Line Ampacity
   Sag & Tension
   HV DC Transmission Link

12) *Transformer:*
   MVA Sizing
   Tap Optimization

13) *DC Systems:*
   Load Flow
   Short-Circuit
   Battery Discharge & Sizing

14) *Control Systems:*
   DC Control Systems Diagram

15*) Data Exchange:*
   DataX
   SmartPlant Electrical Interface
   Microsoft Access & Excel Interface
   e-DPP Interface
   CAD Interface

16) *Real-Time Monitoring & Simulation:*
   Advance Monitoring
   Energy Accounting
   Real-Time Simulation
   Event Playback
   Load Forecasting

17) *Energy Management System:*
   Automatic Generation Control
   Economic Dispatch
   Supervisory Control
   Interchange Scheduling
   Reserve Management

18) Intelligent Load Shedding:
   Load Preservation
   Load Restoration
   Load Shedding Validation

19) *Intelligent Substation:*
   Substation Automation
   Switching Management
   Load Management"

3.2 OBTAINING *ETAP*

ETAP can be obtained directly from the company that owns it, Operation Technology, Inc. or from a regional sales representative. The main office is:

Operation Technology, Inc.
17 Goodyear, Suite 100
Irvine, CA 92618
1-800-477-*ETAP*
1-949-462-0100
fax 1-949-462-0200
http://ETAP.com/

ETAP is sold in modules. These are listed in Section 3.1. Every user buys a base package and then other modules are added as desired. *ETAP* is available in nuclear versions.

ETAP offers a free 60 day Demo version.

All of this section's examples were done with the *ETAP* 6.0 Demo.

3.3 COMPUTER SYSTEM REQUIREMENTS FOR *ETAP* 6.0

Operating System
　Microsoft Windows Vista,
　Microsoft Windows XP (SP2) *Professional* or *Home* Edition,
　or Microsoft Server 2003 (SP2), *Microsoft Server* 2003 R2 (SP2)

Other Program Requirements
　Microsoft Internet Explorer 5.01 or higher (for minimum level as specified by the operating system in use),
　Microsoft.NET Framework v2.0 (SP1),
　or Microsoft.NET Framework v1.1 (SP1)

PC Configuration Requirements
　Parallel port, USB port, or serial port (for stand-alone licensing only)
　CD-ROM or DVD drive
　5 to 80 GB hard disk space (based on project size, number of buses)
　19" monitor recommended (dual monitors highly recommended)
　Minimum display resolution – 1024 x 768

Recommended Hardware Requirements

    6.0.0 Demo

        Intel Pentium 4

        512 MB RAM

    500 Bus Projects

        Intel Pentium 4 – 2.0 GHz (or dual quad core –E6600) or better

        1 GB RAM

    2000 Bus Projects

        (Highly Recommended)

        Intel Pentium 4 – 3.2 GHz with Hyper-Threading Technology (or dual/quad core – E6700) or better with high speed bus or equivalent

        2 GB of RAM (high speed)

    10000 Bus Projects and Higher

        Intel Xeon – 3.2 GHz with Hyper-Threading for dual/quad core – X6800) or better with high speed system bus or AMD equivalent

        2 to 4 GB RAM (high speed)

3.4 *ETAP* DEMO LIMITATIONS

1) It is operational for 60 days. After that, contact Operation Technology to ask for an extension.

2) A maximum of 12 AC buses or 10 DC buses can be evaluated.

3) Circuits and data cannot be saved.

4) On the standard 6.0 Demo CD the active modules are Load Flow, Short-Circuit (ANSI and IEC), *Star*-Protective Device Coordination, and Arc Flash. Most other modules can be made active by requesting Activation Codes from Operation Technology.

5) Printed output reports and plots are restricted to the original example reports included with the Demo.

3.5 *ETAP* TRAINING AND TUTORIALS

ETAP offers free online introductory tutorials, for-a-fee online training, and classroom training. Information on these can be found on the *ETAP* website, http://ETAP.com.

There are 29 free online tutorials. Some are videos, others are text with figures. The beginner should go through these, especially the ones showing how to create one-line diagrams and do load flow studies.

For-a-fee online training is offered by *ETAP*, although at the time this book was composed, most of their for-a-fee training was done in classrooms.

ETAP classroom training is done at their home office and in many locations around the world. Most of these are for a fee.

3.6 SETUP AND INSTALLATION OF THE *ETAP* DEMO

The Demo disk comes with a two page "Setup and Installation Guide". If you do not have the "...Guide", setup is still not difficult. A four-character installation code is needed to setup *ETAP*. If you do not have it, contact Operation Technology.

3.7 EXAMPLE SYSTEM ANALYZED WITH *ETAP*

The same example as was analyzed manually in Section 1.3 is analyzed here with *ETAP* using its Base, Load Flow, and Short-Circuit – ANSI/IEEE Packages.

3.7.1 SETTING UP THE *ETAP* DEMO FOR USE ON THE SYSTEM OF FIGURE 1-3-1

The *ETAP* Demo will do a load flow analysis on a, "New Project". However, using the "New Project" mode with the *ETAP* Demo disables its "Short-Circuit Analysis" capability. To use the "Short-Circuit Analysis" capability of *ETAP* it is necessary to have the Demo bring up its "Example Project (ANSI)" or "Example Project (IEC)" systems. Then those systems are deleted in *ETAP*'s "Edit" mode and are replaced with the desired system.

Here the *ETAP* "Example Project (ANSI)" will be brought up and then modified to become the system of Figure 1-3-1.

1) Start *ETAP* by double left-clicking on the *ETAP* icon.

2) On the "Welcome to *ETAP*" window left-click "OK".

3) On the "Select Demo Project" window select the default "Example Project (ANSI)" and left-click "OK".

4) On the "*ETAP* Logon" window left-click "OK".

5) On the "Select Access Level" window select the default "Project Editor" and left-click "OK".

6) A number of *ETAP* windows will appear. The top and active window is "Study View (Edit Mode)". See Figure 3-7-1-1.

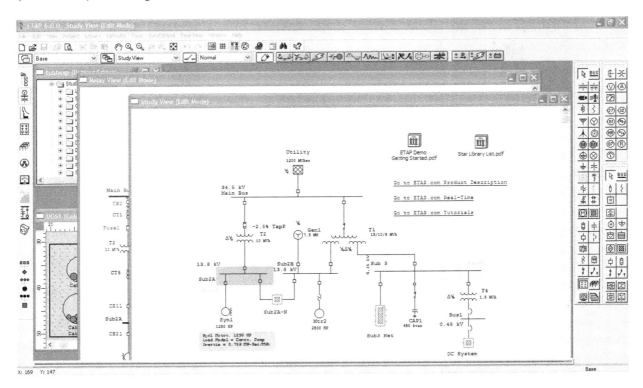

Figure 3-7-1-1 "Example Project (ANSI)" system.

7) Block and delete the system. The "Grid 1" and "Grid 2" colored blocks will be left after the system has been deleted. Drag them to the right side and leave them there. The result can be seen in Figure 3-7-1-2.

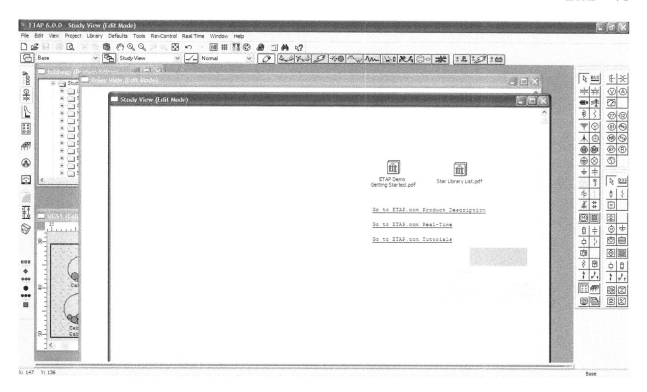

Figure 3-7-1-2 Cleared work window, ready for the Figure 1-3-1 system.

8) On the left of the screen left-click on the "System Dumpster" icon. It looks like a trash can.

9) On the "Dumpster" window delete the "SubOL1" and "SubOL2". This will make designations like T1 and T2 available for use on the system of Figure 1-3-1.

10) The work window is now ready for creating a new one-line diagram.

3.7.2 CREATING A ONE-LINE DIAGRAM

1) The one-line diagram of Figure 1-3-1 will be re-created below with *ETAP*.

2) Left-click on the "Power Grid" icon in the "AC Edit" toolbar (the toolbar near the right side with transformer symbols and "BUS" on it), release the button (do not hold the left mouse button down, as would be done with some CAD programs). Drag the icon to a central location on the work window, then left-click again to set the icon in place. After an icon is placed on the one-line diagram it can be moved by left-clicking on it, holding the left button down, and then releasing the left button when the icon is in the proper location. See Figure 3-7-2-1.

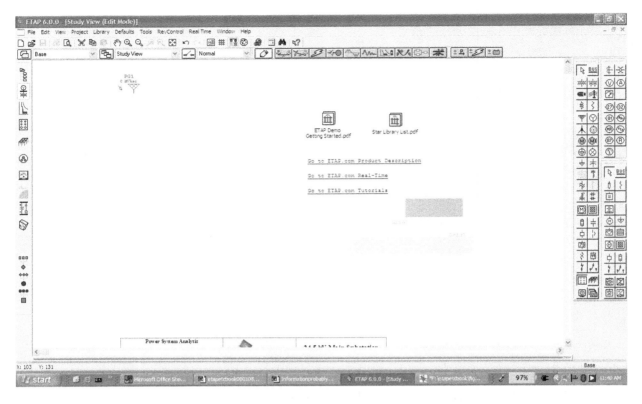

Figure 3-7-2-1 Work window with "Power Grid" icon in place.

3) Left-click, drag, and set a "BUS" icon beneath the terminal of the work window's "Power Grid" icon.

4) With the cursor on the terminal of the "Power Grid" icon, left-click and hold down the mouse button. The terminal will turn purple. Drag a line straight down to the "BUS" icon until the "BUS" icon turns purple. Release the left mouse button and the "Power Grid" icon will be connected to the "BUS" icon by a red line.

5) Left-click and drag the other icons to the work window and then connect them. When needed, the "BUS" icon can be stretched by left-clicking on an end and dragging it. The result is shown in Figure 3-7-2-2.

Note:
1) *ETAP* will not allow two impedance components in series without a "BUS" or "NODE" between them. Cables and transformers are examples of impedance components.
2) *ETAP* will allow two or more zero impedance components in series without interconnecting "BUS"es or "NODE"s. Fuses, circuit breakers, and switches are examples of zero impedance components.
3) *ETAP* can use "BUS"es or "NODE"s between series components. "BUS"es can accept many components. "NODE"s can only accept two components.

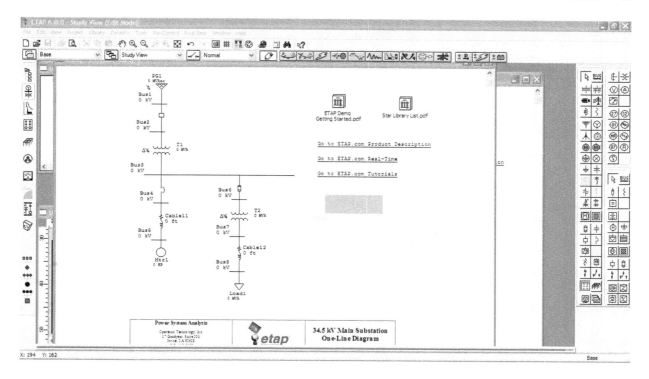

Figure 3-7-2-2 Work window of the Figure 1-3-1 example one-line diagram with default transformer types. No system data has been entered yet.

6) Note that the cables are labeled "Cable 11" and "Cable 12" rather than "Cable 1" and "Cable 2". This is because the lower cable numbers are still saved by *ETAP* for the original *ETAP* Demo example.

3.7.3 LOAD FLOW ANALYSIS

3.7.3.1 Circuit Data Entry

ETAP has many data windows for each component. With these, the user can input a great amount of data. However, most of this data is not needed by *ETAP* for load flow and balanced short-circuit analyses. Here, only the required data will be entered.

Underlined values are entered into *ETAP*.

3.7.3.1.1 "Power Grid Editor – PG1" data

1) Right-click on the "Power Grid" icon and select "Properties" to bring up the "Power Grid Editor – PG1" "Info" tab shown in Figure 3-7-3-1-1-1.

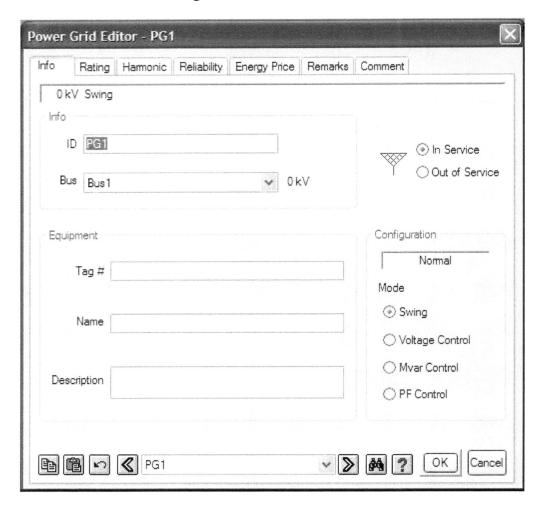

Figure 3-7-3-1-1-1 "Power Grid Editor" "Info" tab.

2) Values needed for the "Power Grid Editor – PG1" "Rating" tab are taken from the manual analysis information in Sections 1.3.1.2.1 and 1.3.1.2.2. They are:

Rated kV = <u>3.8</u>
Positive sequence RS1 = .2 Ω
Positive sequence XS1 = .3 Ω

The underlined value, <u>3.8</u>, can be entered directly into the "Power Grid Editor – PG1" "Rating" window. The impedances need to be converted to per cent values with a 100 MVA base.

The values that *ETAP* will accept are calculated below and underlined:

$$\%RPOS = RS1 \times (\text{Power base}/VLL^2) \times 100$$
$$= .2 \times (100 \times 10^6/3800^2) \times 100 = \underline{138.5}$$

$$\%XPOS = XS1 \times (\text{Power base}/VLL^2) \times 100$$
$$= .3 \ (100 \times 10^6/3800^2) \times 100 = \underline{207.8}$$

3) For load flow analysis and balanced short-circuit analysis, values for the negative and zero sequence impedances and the connection configuration are not needed. Here, those impedances and the connection configuration will be left at default values.

4) Select the "Rating" tab and enter 3.8, 138.5, and 207.8 into the "Power Grid Editor – PG1". Also enter 100 for the "Operating %V". This shows that 100% of the 3.8 kV is applied to the system. The filled out "Rating" tab is shown in Figure 3-7-3-1-1-2.

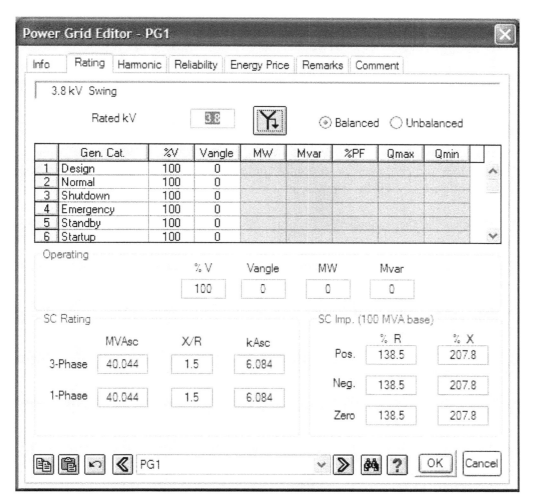

Figure 3-7-3-1-1-2 "Power Grid Editor –PG1" "Rating" tab.

3.7.3.1.2 Circuit breaker C1 and C2 and fuse F1 data

No data input is necessary.

3.7.3.1.3 Transformer T1 data

1) Transformer T1 data is taken from the manual analysis information in Section 1.3.1.1.1.

Rating Prim. kV = <u>3.8</u>
Rating Sec. kV = <u>0.48</u>
Rating MVA = <u>.6</u>
The transformer % impedances (.6 MVA and 3800/480 V rms base) are:
  Positive sequence %RT11 = 1.73%
  Positive sequence %XT11 = 1.52%
These need to be converted to units that *ETAP* will accept.

$$\text{Impedance Positive } \%Z = (\%RT11^2 + \%XT11^2)^{1/2}$$
$$= (1.73^2 + 1.52^2)^{1/2} = \underline{2.30}$$

$$\text{Impedance Positive } X/R = \%XT11/\%RT11$$
$$= 1.52/1.73 = \underline{.879}$$

2) For load flow analysis and balanced short-circuit analysis, values for the negative and zero sequence impedances and the connection configuration are not needed. Here, those impedances and the connection configuration will be left at default values.

3) Right-click on the transformer T1 icon, select "Properties" and the "Rating" tab, and enter the underlined data to create the "2-Winding Transformer Editor – T1" window shown in Figure 3-7-3-1-3-1.

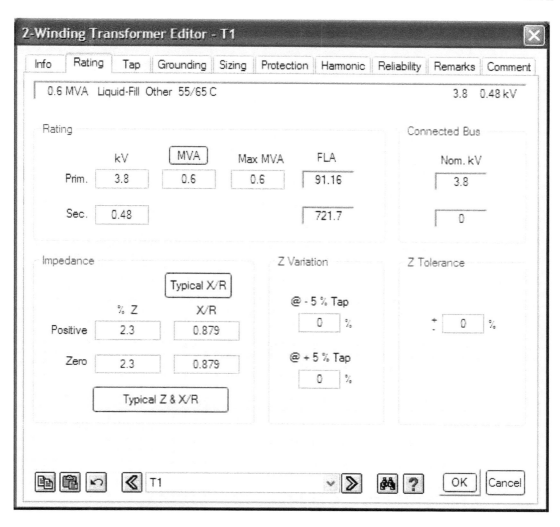

Figure 3-7-3-1-3-1 Transformer "2 Winding Transformer Editor – T1" "Rating" tab window.

3.7.3.1.4 Transformer T2 data

1) Transformer T2 data is taken from the manual analysis information in Section 1.3.1.1.2.

Rating Prim. kV = .48
Rating Sec. kV = 0.12
Rating MVA = .09
The transformer % impedances (.09 MVA and 480/120 V rms base) are:
Positive sequence %RT11 = 2.00%
Positive sequence %XT11 = 1.80%
These need to be converted to units that *ETAP* will accept.

$$\text{Impedance Positive } \%Z = (\%RT11^2 + \%XT11^2)^{1/2}$$
$$= (2.00^2 + 1.80^2)^{1/2} = 2.69$$

$$\text{Impedance Positive } X/R = \%XT11/\%RT11$$
$$= 1.80/2.00 = .900$$

2) The transformer data is entered as it was in Section 3.7.3.1.3.

3.7.3.1.5 Data for the cable from circuit breaker C2 to motor M, "Cable 11"

1) From the manual analysis information in Section 1.3.1.3.3:

Length = 200 ft
Impedance (per conductor) Pos. R = 0.078, (Ω/1000 ft.)
Impedance (per conductor) Pos. X = 0.030, (Ω/1000 ft.)
Select Units Ω per 1000 ft

2) For load flow analysis and balanced short-circuit analysis, values for the negative and zero sequence impedances and the connection configuration are not needed. Here, those impedances and the connection configuration will be left at default values.

3) Two Cable Editor windows are needed to use the raw data of Section 1.3.1.3.3, the "Info" tab window and the "Impedance" tab window. Right-click on the cable icon between the circuit breaker C2 and the motor M to bring up the Cable Editor windows. The filled out windows are shown in Figures 3-7-3-1-5-1 and 3-7-3-1-5-2.

Figure 3-7-3-1-5-1 "Cable Editor – Cable 11" "Info" tab window.

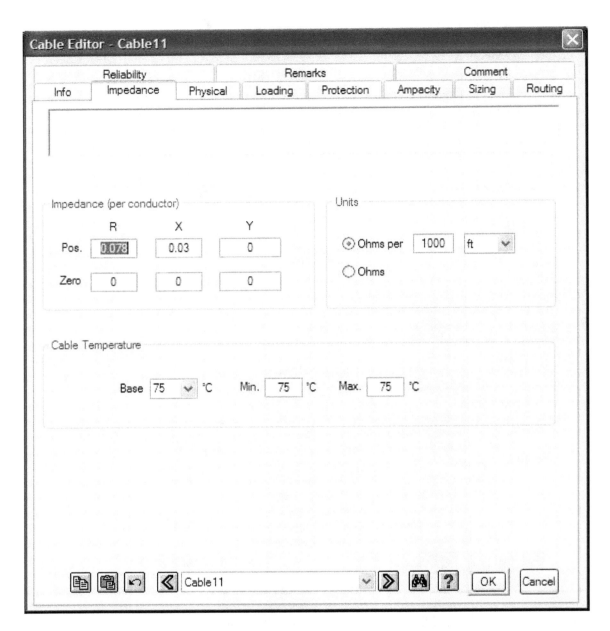

Figure 3-7-3-1-5-2 2-"Cable Editor – Cable 11" "Impedance" tab window.

3.7.3.1.6 Data for the cable from transformer T2 to static load R, "Cable 12"

    1) From the manual analysis information in Section 1.3.1.3.4:

        Length = <u>200</u> <u>ft</u>
        Impedance (per conductor) Pos. R = <u>0.098</u>, (Ω/1000 ft.)
        Impedance (per conductor) Pos. X = <u>0.030</u>, (Ω/1000 ft.)
        Select Units <u>Ω per</u> <u>1000</u> <u>ft</u>

    2) The cable data is entered here as it was in 3.7.3.1.5

3.7.3.1.7 Motor M data

    1) Motor M data is taken from the manual analysis information in Section 1.3.1.4.

        Ratings HP = <u>250</u>
        Ratings kV = <u>0.48</u>

    3) Also select <u>MFR</u> for typical manufacturer's data relating to the 250 hp and .48 kV.

    2) The data is entered on two windows the "Nameplate" window and the "Typical Nameplate Data" sub-window. See Figures 3-7-3-1-7-1 and 3-7-3-1-7-2.

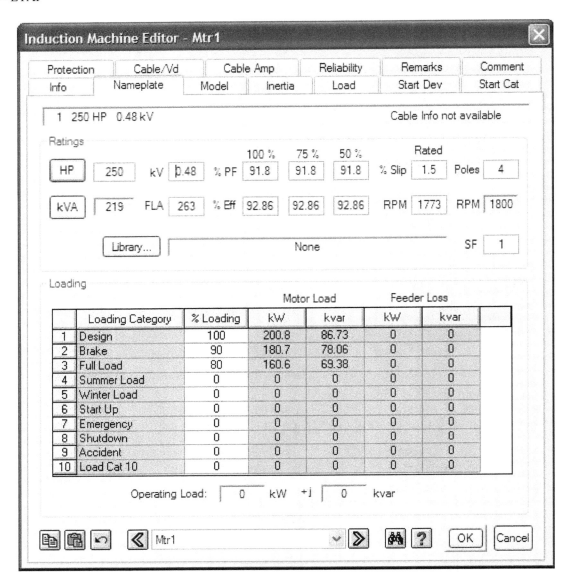

Figure 3-7-3-1-7-1 Motor M "Induction Machine Editor Mtr1" "Nameplate" tab window.

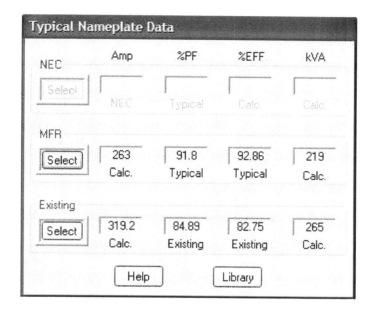

Figure 3-7-3-1-7-2 Motor M "Typical Nameplate Data", sub-window of the "Nameplate" tab window.

3.7.3.1.8 "Static Load" R data

1) "Static Load" R data is taken from the manual analysis information in Section 1.3.1.5.

Ratings kV = 0.12
Ratings MVA = 0.030

2) For load flow analysis and balanced short-circuit analysis, values for the negative and zero sequence impedances and the connection configuration are not needed. Here, those impedances and the connection configuration will be left at default values.

3) Right-click on the "Static Load" icon, the "Loading" tab, and then enter the underlined data to create the "Static Load Editor – Load1" window shown in Figure 3-7-3-1-8-1.

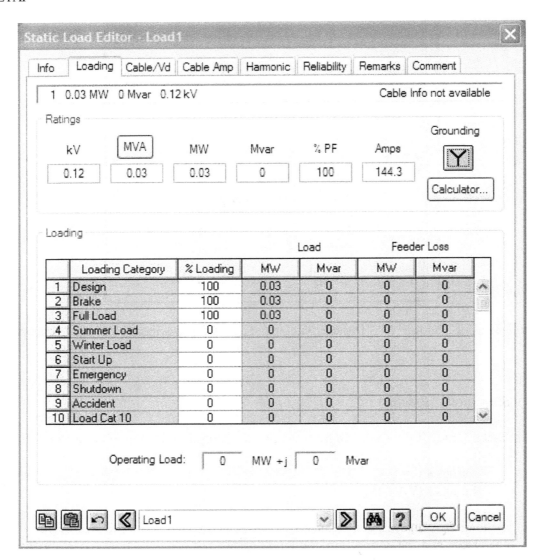

Figure 3-7-3-1-8-1 R "Static Load Editor Load1" "Loading" tab window.

3.7.3.2 Load Flow Analysis

1) Left-click on the upper icon that looks like a balanced scale with a P and Q on it. This is next to the "Edit" icon and is titled "Load Flow Analysis".

2) On the toolbar on the right left-click the icon that has p kV and q on it. This is titled "Run Load Flow".

3) The work window will be replaced by a "Load Flow Analysis" window. By default it contains line currents (in A rms) and percent voltages. See Figure 3-7-3-2-1.

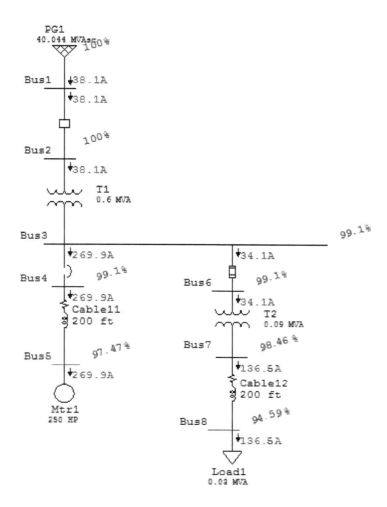

Figure 3-7-3-2-1 Default window of "Load Flow Analysis".

4) The output can be customized by using the selection tool on the right toolbar. Its icon looks like a computer monitor and it is titled "Display Options". Select the "Voltage Unit" to be "V" to display bus voltages. This will make it easier to compare the *ETAP* results with those of the manual analysis. See Figure 3-7-3-2-2 for "Load Flow Analysis" with "V" as the "Voltage Unit".

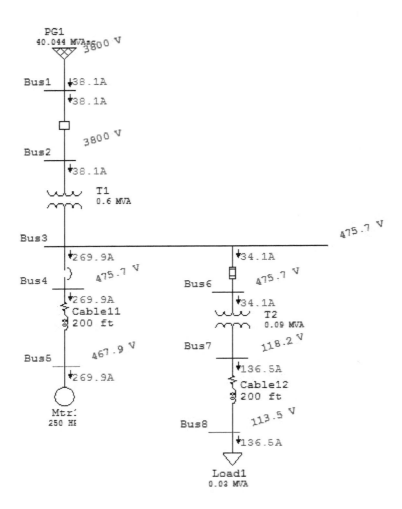

Figure 3-7-3-2-2 "Load Flow Analysis" output with bus voltages shown.

3.7.4 BALANCED SHORT-CIRCUIT ANALYSIS

1) Left-click on the upper icon that looks like a lightning bolt. This icon is titled "Short-Circuit Analysis".

2) The short-circuit location must be designated. To do this left-click on the icon above and to the right of the lightning bolt icon. The icon looks like a suitcase and is titled, "Edit Case Study".

3) A window titled, "Short Circuit Case Study" opens. On this select "Bus 2", the bus just downstream of circuit breaker C1. See Figure 3-7-4-1.

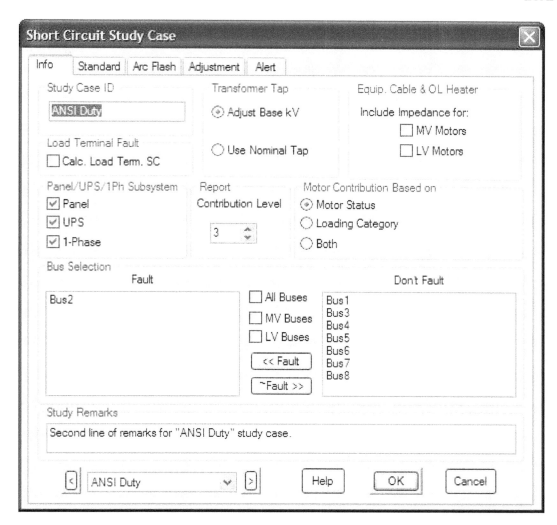

Figure 3-7-4-1 Short Circuit Case Study window, "Bus 2" selected.

4) Left-click on the icon on the right hand side of the screen, the one that has three lines going into one and has "duty" written under it. This icon is titled "Running 3-Phase Device Duty (ANSI 37)". The result is shown in Figure 3-7-4-2.

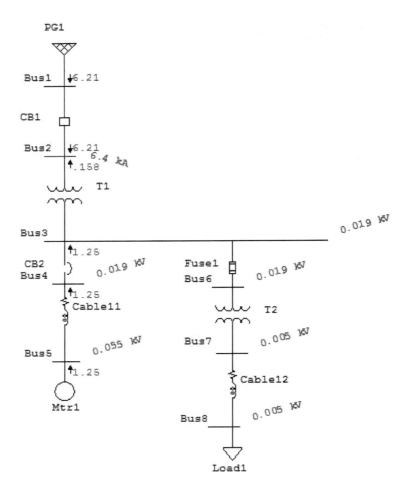

Figure 3-7-4-2 Currents (in kA) during a balanced short circuit at "Bus 2", the output of circuit breaker C1.

5) In a similar way, the balanced short-circuit currents can be found at "Bus 4" and "Bus 6". See Figures 3-7-4-3 and 3-7-4-4.

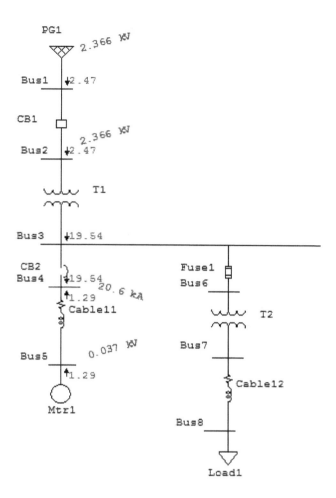

Figure 3-7-4-3 Currents (in kA) during a balanced short circuit at "Bus 4", the output of circuit breaker C2.

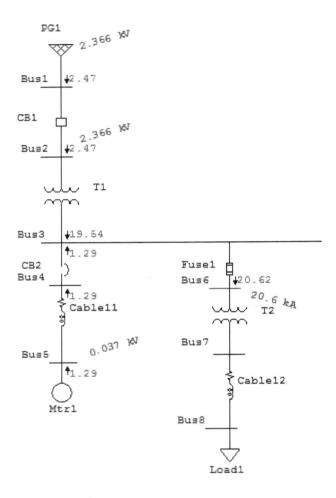

Figure 3-7-4-4 Currents (in kA) during a balanced short circuit at "Bus 6", the output of fuse F1.

3.7.5 UNBALANCED SHORT-CIRCUIT ANALYSIS

Unbalanced short-circuit analysis requires more information than load flow and balanced short-circuit analysis require. The voltage supply and the impedances used in the load flow and balanced short-circuit analyses are *symmetrical components* positive sequence values. For unbalanced short-circuit analysis *symmetrical components* negative and zero sequence impedances are also needed. Furthermore, unlike balanced analyses, the *ETAP* one-line diagram has to have correct transformer connections (i.e. Y-Y, Y-Y(ground), Y-Δ, etc.).

3.7.5.1 "Power Grid Editor PG1" Data

1) *ETAP* has automatically selected source and line negative and zero sequence impedances that are equal to their positive sequence impedances. This is correct for the negative sequence impedances, but the zero sequence impedances must be changed, to those of Section 1.3.1.2.2 of the manual calculation. Right-click on the "Power Grid" icon, select "Properties", and bring up the "Rating" tab shown in Figure 3-7-5-1-1.

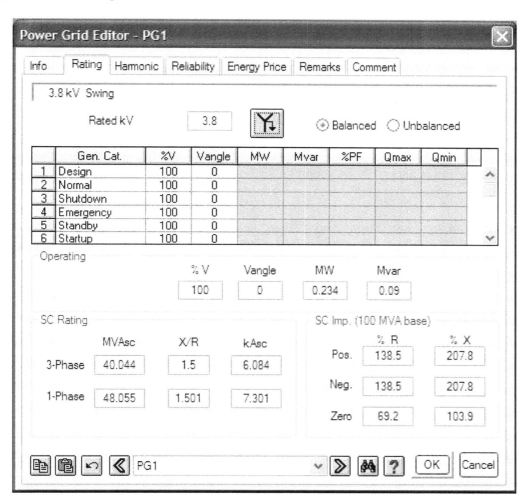

Figure 3-7-5-1-1 "Power Grid Editor – PG1" "Rating" tab with correct zero sequence impedances.

2) Enter zero sequence impedances to be half the value of the positive sequence impedances, as they were with the manual calculation values. The entered values should be:

$$\%RZERO = (138.5\%)/2 = \underline{69.2}$$

$$\%XZERO = (207.8\%)/2 = \underline{103.9}$$

3.7.5.2 Circuit Breaker C1 and C2 and Fuse F1 Data

No data input is necessary.

3.7.5.3 Transformer T1 Data

1) *ETAP* automatically makes the negative and zero sequence impedances the same as the positive sequence impedances, as they are in Section 1.3.1.1.1. *ETAP* allows the zero sequence impedances to be changed, but that is not necessary in this example.

2) The winding connections and "Solid" neutral grounding of the transformer should be selected with the "Grounding" tab window. The "Grounding" tab window is shown in Figure 3-7-5-3-1.

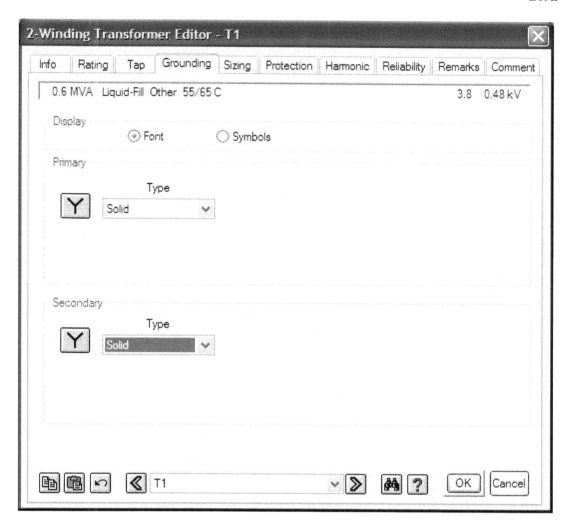

Figure 3-7-5-3-1 Transformer T1 "Grounding" tab window.

3.7.5.4 Transformer T2 Data

1) *ETAP* automatically makes the negative and zero sequence impedances the same as the positive sequence impedances, as they are in Section 1.3.1.1.2. *ETAP* allows the zero sequence impedances to be changed, but that is not necessary in this example.

2) The winding connections of the transformer should be selected with the "Grounding" tab. The "Grounding" tab window is shown in Figure 3-7-5-4-1.

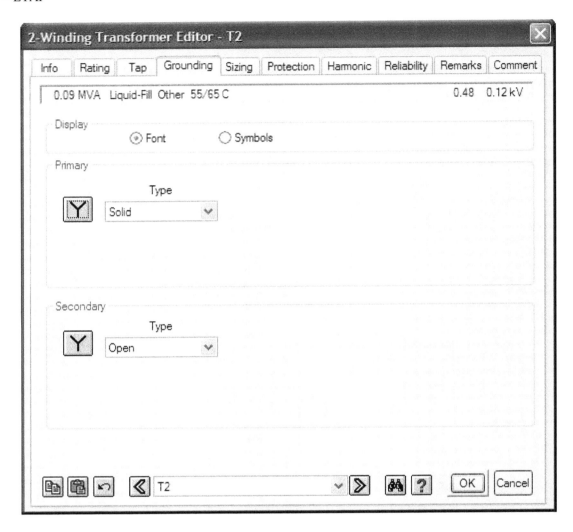

Figure 3-7-5-4-1 Transformer T2 "Grounding" tab window.

3.7.5.5 Data for the Cable from Circuit Breaker C2 to Motor M, "Cable 11"

ETAP assumes that the negative sequence impedances are the same as the positive. The zero sequence impedances need to be entered manually. In Section 1.3.1.3.3 the zero sequence impedances are the same as the positive. They should be entered into the "Impedance" tab window as shown in Figure 3-7-5-5-1.

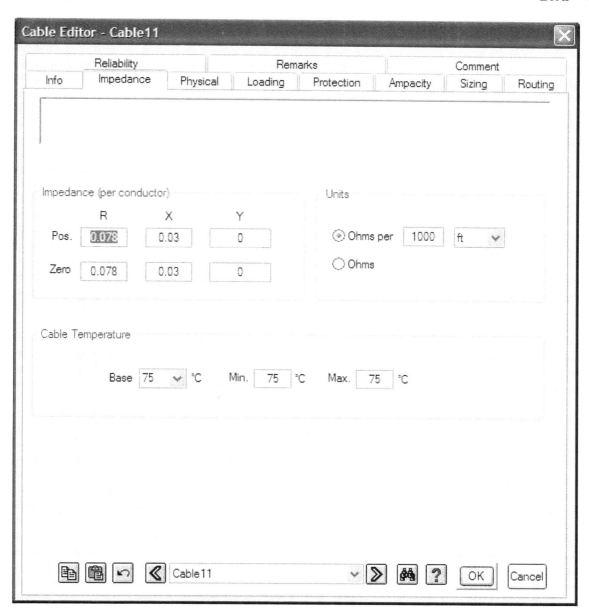

Figure 3-7-5-5-1 2-Cable Editor "Impedance" tab window, "Cable 11".

3.7.5.6 Data for the Cable from Transformer T2 to Static Load R, "Cable 12"

ETAP assumes that the negative sequence impedances are the same as the positive. The zero sequence impedances need to be entered manually. In Section 1.3.1.3.4 the zero sequence impedances are the same as the positive. They should be entered into the "Impedance" tab window as they were for "Cable 11" in Section 3.7.5.5.

3.7.5.7 Motor M Data

ETAP selected typical negative and zero sequence impedances when motor data was input. It is not necessary to add further motor data to the *ETAP* program.

3.7.5.8 Static Load R Data

ETAP assumes that the negative sequence impedances are the same as the positive. The zero sequence impedances are infinite, since the neutral is isolated. It is not necessary to add further static load data to the *ETAP* program.

3.7.5.9 Unbalanced Short-Circuit Analysis of a Line-to-Ground Short Circuit

1) Left-click on the upper icon that looks like a lightning bolt. This icon is titled "Short-Circuit Analysis".

2) The short-circuit location must be designated. To do this left-click on the icon above and to the right of the lightning bolt icon. The icon looks like a suitcase and is titled, "Edit Case Study".

3) A window titled, "Short Circuit Case Study" opens. On this select 'Bus 5", the bus just upstream of motor M.

4) Just to the left of the suitcase icon there is a small window that allows the selection of the "Study Case". If it has not already been selected, select "ANSI - LG".

5) Select the "Run 3-Phase, LG, LL, LLG(1.5 – 4 Cycle)" icon in the "Short-Circuit Toolbar". The icon looks like a lightning bolt with a number 4 and a sine wave under it. This will predict the currents and voltages within the time from 1.5 to 4 cycles from the start of the short circuit.

6) On the "Short-Circuit Toolbar" on the right select the "Display Options Short-Circuit" icon, the one that looks like a computer monitor. In the window that comes up select in "Fault Type", "L-G", and "Phase Values (A, B, C)". See Figure 3-7-5-9-1. Then left-click "OK". The resulting display is shown in Figure 3-7-5-9-2.

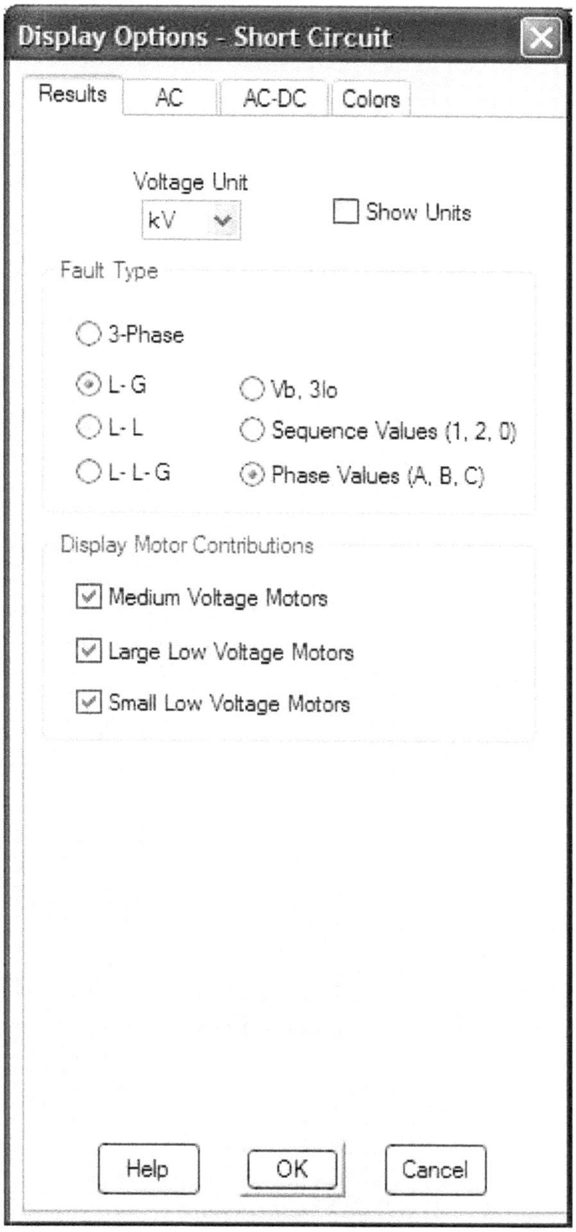

Figure 3-7-5-9-1 "Display Options Short-Circuit" window.

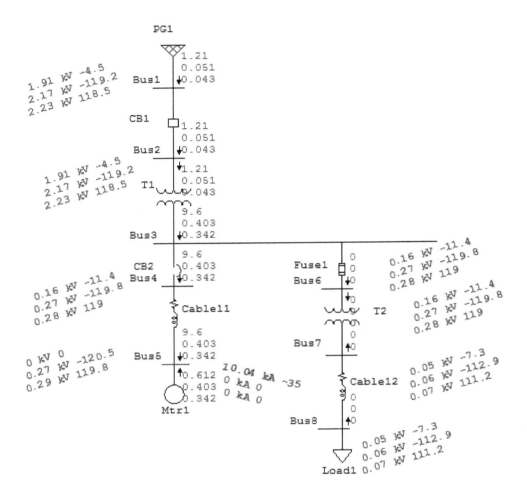

Figure 3-7-5-9-2 Line-to-ground short circuit at "BUS 5", the input to the motor M. Phase voltages and currents (in kA) are displayed. Values have been dragged to make reading easier.

3.8 *ETAP* REFERENCE

Operation Technology, Inc., *ETAP 6.0 Demo Getting Started*, 2006. It is 173 pages long and 8.5" x 11". A person learning *ETAP* should get a copy. It is available for free on the *ETAP* website.

4.0 *EDSA*

The *EDSA Micro Corporation* is also simply called *EDSA*. Originally, *EDSA* was an acronym for **E**lectrical **D**esign **S**imulation & **A**nalysis. *EDSA* develops program solutions for electrical power systems. Their *Paladin* program products have been used for more than 25 years.

This book is concerned with their *Paladin DesignBase* Version 2.0 electrical power system analysis program. In the electrical industry, the *Paladin DesignBase* program is sometimes referred to by its company's name, *EDSA*.

4.1 FEATURES, AS STATED BY *EDSA*

EDSA states:

"*Paladin DesignBase* is an internationalized, ubiquitous modeling system that can be used to create, analyze and simulate virtually any type of electrical power distribution system regardless of complexity. It provides the technological richness needed to analyze systems from a variety of perspectives: from static to dynamic simulations, including the ability to model and embed the detailed control logic of the intelligent electronic devices responsible for controlling how power flows throughout the system.

EDSA products meet safety standards established by the U.S. Nuclear Regulatory Commission (NRC) and the Nuclear Procurement Issues Committee (NUPIC). *Paladin DesignBase* complies with IEEE/ANSI, IEC and Mil Standards.

In addition, a library of more than 50 optional *Paladin DesignBase* Solutions Modules allow users to perform specialized forms of analysis and optimization, including Fault Analysis, Protection Coordination, Power Flow Analysis, Power Quality Analysis and Mitigation, Dynamic Behavior Simulation, Design Optimization, and Sizing Optimization.

Paladin DesignBase provides three crucial benefits for users:

The *DesignBase* serves as the basis for powerful modeling, simulation, analysis, and 'what if' testing in a teaming environment, and in an integrated, easy-to-use modeling environment.

The engineering intent behind all facets of the electrical engineer's design is unambiguously documented and preserved... to guide downstream construction, installation, and maintenance crews.

Once the facility is operational, the *DesignBase* can – using *EDSA*'s *Paladin Live* platform – serve as benchmark against which actual measurements are compared, in order to identify potential operational problems while they are still in the formative stages... and there is still time to preempt them."

Typically *Paladin DesignBase* users purchase its modules in packages. Users buy a base module and some other modules to create their own total program package. The packages may contain from 3 to over 50 modules. Below is a copy of the 2009 *EDSA* website list of *Paladin DesignBase* modules:

"1) *Base Modules (included with all package types)*
    Editor (Graphical Interface)
    Loader
    Build Devices

2) *Fault Analysis*
    AC Short Circuit – 3 Phase ANSI/IEEE & PDE and IEC 909
    AC Single Phase Short Circuit
    AC & DC Arc Flash Evaluation (NFPA 70E 2004 & IEEE 1584)
    AC Short Circuit – IEC 363
    DC Short Circuit Classical & IEC 61660

3) *Protection Coordination*
    AC & DC System Protective Device Coordination
    Distance Relays

4) *Power Flow Analysis*
    Advanced Power Flow
    Advanced Single Phase Power Flow
    Advanced Motor Starting
    Power Flow Phase Development
    AC Voltage Profile
    Integrated AC/DC Load Flow
    Object Orientated DC Load Flow
    PSO (Power System Optimization)
    Voltage Stability & Contingency Analysis
    Load Forecasting

5) *Power Quality Analysis and Mitigation*
      Harmonics Analysis (with Automatic Filter Sizing)
      Reliability Worth Assessment of Distribution System
      Capacitor Sizing

6) *Dynamic Behavior Simulation*
      Advanced Transient Stability Analysis (with Model Builder)
      EMTAP (Electromagnetic Transient Analysis)

7) *Design Optimization*
      Schedules
      Induction Motor Parameters
      Computation of Impedance Parameters – Sync Motors
      Advanced Substation Grounding Grid Design 3-D Plotting
      Cable Pulling
      Cable Magnetic Field
      Transmission Line Parameters
      Shielding
      Advanced Power Cable Ampacity

8) *Sizing Optimization*
      Battery Sizing
      NEC Wire Sizing
      Generator Set Sizing
      Bare wire Sizing
      Short Line Parameters
      Sag and Tension"

4.2 OBTAINING *PALADIN DESIGNBASE*

Paladin DesignBase can be obtained directly from *EDSA,* any of its 21 international agents or its U.S. manufacturers representatives as listed on their web site. *EDSA*'s U.S. headquarters are:

EDSA Micro Corporation
16870 West Bernardo Drive, Suite 330
San Diego, CA 92127
1-800-362-0603
http://EDSA.com/

4.3 COMPUTER SYSTEM REQUIREMENTS FOR *PALADIN DESIGNBASE* 2.0

Operating System
      Microsoft Windows Vista,
      Microsoft Windows XP,
      or *Microsoft Windows* 2000

Other Programs
      Microsoft Internet Explorer
      Microsoft Office 2000 or higher

PC Configuration
      CD-ROM drive
      160 GB available hard disk space (3.8 GB is satisfactory for analyzing small systems)
      Mouse/pointer (*Microsoft IntelliMouse* recommended)
      Monitor with at least 1024 x 768 display resolution (24 bit recommended)
      Sound card and speakers (optional for tutorials)

Recommended Hardware Requirements
      Intel Pentium 4 or *AMD Athlon XP* or higher processor
      2 to 4 GB RAM (.5 GB is satisfactory for analyzing small systems)

4.4 *PALADIN DESIGNBASE* TRIAL VERSION LIMITATIONS

The Trial Version is a full-featured version capable of handling 100 buses with all the options that come with the "PE"-type package. Normally, the Trial Version is active for only 14 days. If more time is needed, contact *EDSA* to ask for an extension.

4.5 *PALADIN DESIGNBASE* TRAINING AND TUTORIALS

EDSA offers *Paladin DesignBase* CDs with free introductory video tutorials (playable on a computer with *Windows*), guides, and manuals. There are 57 separate video tutorials on one CD and 53 written guides and manuals on another. These CDs accompany the Trial Version CD. They also offer for-a-fee classroom training at their San Diego headquarters; this three day training comes free for one user with any software purchase. *EDSA* training courses qualify for Continuing Education Units (CEUs) and Professional Development Hours (PDHs). IEEE members also qualify for a 10% discount on normal course tuition. Further information on these can be found on the *EDSA* website, http://www.edsa.com/.

EDSA also provides several interactive demos of their software's basic features on their web site.

The beginner should go through the introductory video tutorials, guides, and manuals, especially the ones showing how to create a one-line diagram, do power-flow (called load flow by *SKM* and *ETAP*) analysis, and do short-circuit (fault) analysis.

For the beginner, the most useful video tutorials are: *Building 3phase One Line*, *Plug-Sockets*, *Back Annotation*, *Power Flow Simulation*, and *Fault Analysis All Buses*.

The document guides and manuals are much more detailed than the videos. For the beginner, the most useful of them are: *Getting Started with Paladin DesignBase* <file 1> (main.pdf) and *Getting Started with Paladin DesignBase* <file 2> (users_guides_main.pdf). These two link to other useful documents on the CD such as, *Quick Start Guide* and *Creating Your Job File in DesignBase*. Some of the other documents on the CD that would be of interest to the beginner are *Complete Installation Guide* (Installation_Guide.pdf), and *Short Circuit Analysis Program ANSI/IEC/IEEE & Protective Device Evaluation* (3_Phase_Short_Circuit.pdf).

4.6 SETUP AND INSTALLATION OF THE *PALADIN DESIGNBASE* TRIAL VERSION

The *Paladin DesignBase* program disk comes with a printed 42 page *Installation Guide*. There is also a 57 page *Installation Guide* on the *Documents* CD. On the CD it can be accessed through *Getting Started with Paladin DesignBase* <file 1> (main.pdf) or directly from Installation_Guide.pdf.

If only 14 days of activation are needed, then it is possible to install the Trial Version without contacting *EDSA*. Just follow the procedures given in the *Installation Guides*, but type nothing in when the computer asks for a User Name, Organization, and Serial Number. Also, do not select "Configure EDSA Client License" or "Create EDSA Admin Account". Finally in the last window, the "Paladin Client Licensure Manager V2.00.00" window, just select "Close". The computer should install the program so that it can be used for 14 days.

If it is anticipated that longer than 14 days of activation are needed, an *EDSA* sales representative should be contacted in advance of installing the program. In some cases, longer trial periods are allowed. To have a longer trial period, *EDSA* would provide you with a temporary serial number and password.

4.7 EXAMPLE SYSTEM ANALYZED WITH *PALADIN DESIGNBASE*

The same example as was analyzed manually in Section 1.3 is analyzed here with *Paladin DesignBase* using its power-flow and short-circuit capabilities.

4.7.1 CREATING A JOB FILE AND DRAWING FILE FOR USE ON THE SYSTEM OF FIGURE 1-3-1

1) Start *Paladin DesignBase* by double left-clicking on its icon. The window shown in Figure 4-7-1-1 appears.

Figure 4-7-1-1 Opening window of *Paladin DesignBase*.

2) On the opening window, left-click on "File", and then on "New Drawing File…" A window titled "New Drawing" will appear. Select the "*EDSA*" tab in that window.

3) On the "*EDSA*" window, left-click on "Electrical One-Line AC 3Phase.axt". See Figure 4-7-1-2.

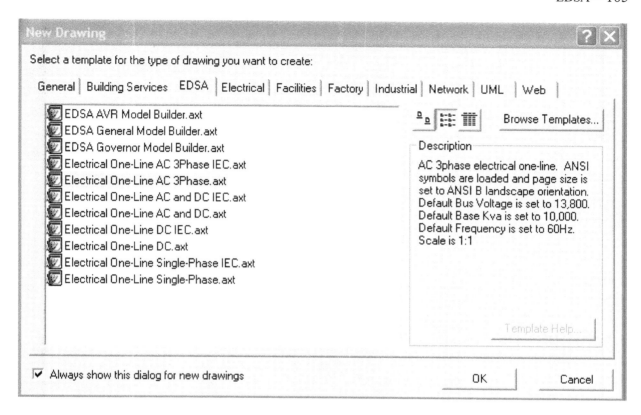

Figure 4-7-1-2 Window selecting "Electrical One-Line AC 3Phase.axt".

4) Left-click on "OK". A "Create a New *EDSA* Project" window will appear. In that window type in the file name "exampleEDSA". The directory where the file is to be saved could also be selected. Here the default location will be used. See Figure 4-7-1-3.

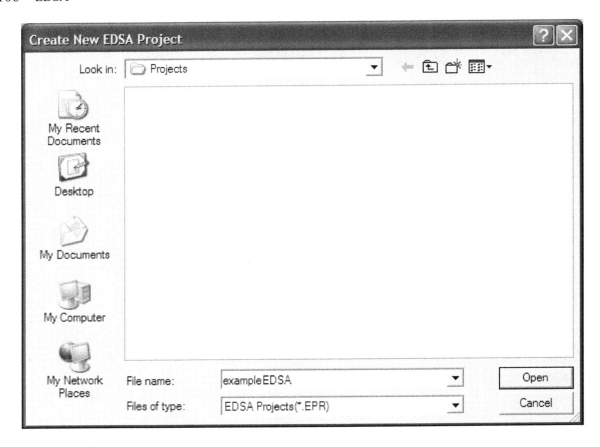

Figure 4-7-1-3 Window to save "exampleEDSA" file.

5) Left-click on "Open". On the "Select Access Level" window select the default "Project Editor" and left-click "OK". A "Data Protection" window appears. The use of a password is optional. Then left-click on "OK". An "EDSA Edit JobFile [example EDSA] Master File" window appears. This is in Figure 4-7-1-4.

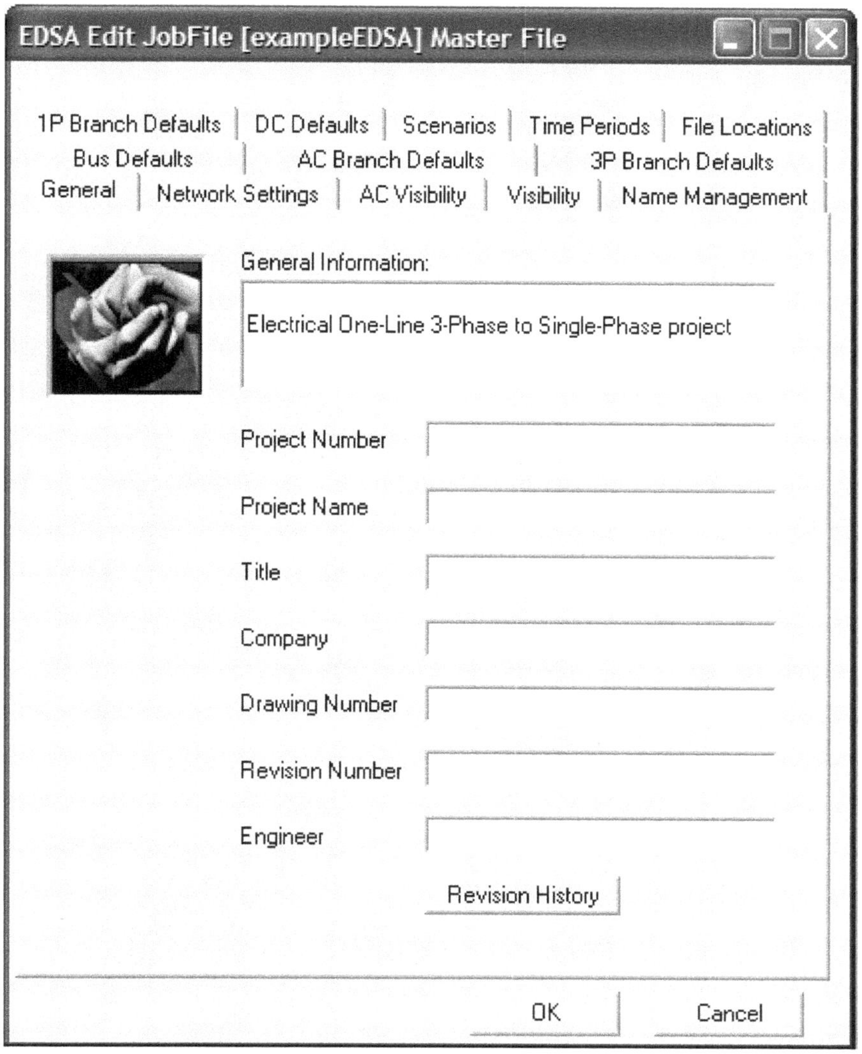

Figure 4-7-1-4 "EDSA Edit JobFile [example EDSA] Master File" window for entering "JobFile" data.

6) The information requested in the "EDSA Edit JobFile [example EDSA] Master File" window is useful for projects that need to be stored. However, for the "exampleEDSA" the fields can be left blank and "OK" left-clicked.

7) The blank work window of Figure 4-7-1-5 appears.

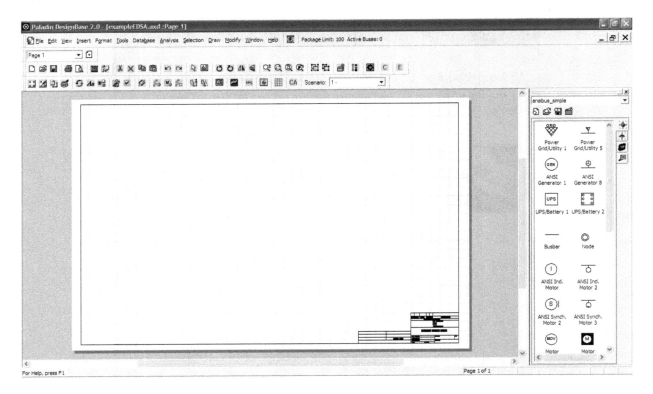

Figure 4-7-1-5 Blank work window for "exampleEDSA", ready for the Figure 1-3-1 system.

4.7.2 CREATING A ONE-LINE DIAGRAM

1) The one-line diagram of Figure 1-3-1 will be re-created below with *Paladin DesignBase*.

2) Left-click and hold the button down on the "Power Grid/Utility 1" icon in the "ansibus_simple" toolbar (the toolbar on the right side), drag the icon to a central location on the work window, and then release the button to set the icon in place. After an icon is placed on the one-line diagram, then it can then be moved by left-clicking on it, holding the left button down, and then releasing the left button when the icon is in the proper location. The "Power Grid/Utility 1" icon may be very small. To increase its size, zoom-in using "Zoom In/Out" under the "View" menu. See Figure 4-7-2-1.

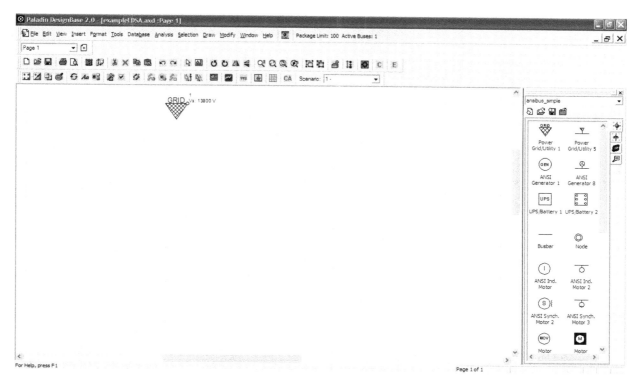

Figure 4-7-2-1 Work window with the "Power Grid/Utility 1" zoomed-in icon in place.

3) Save the drawing file by left-clicking on "File" and "Save Drawing File".

4) Set up the "AutoSnap" feature so that the icons will easily snap together. To do this, go to "Tools" and then select "AutoSnap". On the "AutoSnap" window, left-click on "Clear All", and then select "Sockets" and "Geometry", as shown in Figure 4-7-2-2.

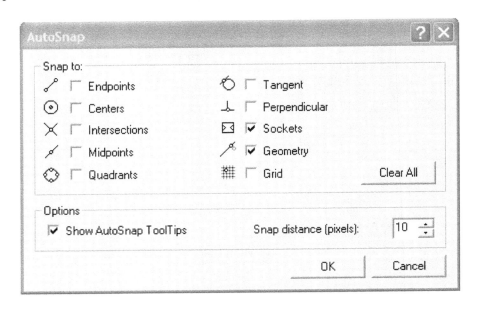

Figure 4-7-2-2 "AutoSnap" window.

5) The icons useful for creating three-phase one-line diagrams are on the "ansibus_simple" and "ansibra_simple" toolbars. Left-clicking the icons to the right of the icon toolbar will switch from one toolbar to another.

6) Left-click and drag the rest of the circuit's icons to the work window. Connect the icons by touching them to the icon they are to connect to and then releasing the left mouse button. When an icon being added is properly positioned to connect, a message, "Connect to Socket" or "Connect to Geometry", will appear near the connection point. If an icon is properly connected to a live icon, it appears black. If it appears as another color, then it is not connected.

7) To insert a bus bar, first set the bus bar in a place near the icon it is to connect to. Then drag the ends of the bus bar out to stretch it to a desired width. Next left-click on the desired icon to select it. Finally, drag and drop the desired icon's lead on to the bus bar. When the dragged lead is ready to be connected a message "Connect to Geometry" appears nearby. Then, the left mouse button can be released. When the "BUS" is properly connected to a live lead it becomes black.

8) The "EDSA Tools" "Back Annotation" window is used to select the values that will be displayed next to the icons. To get to the "Back Annotation" window, left-click on the "Back Annotation" icon, the icon that has an "Aa" in it. This is in the lower row of the upper icons, sixth to the right from the toolbar left. The selected annotations are shown in the windows in Figures 4-7-2-3 and 4-7-2-4.

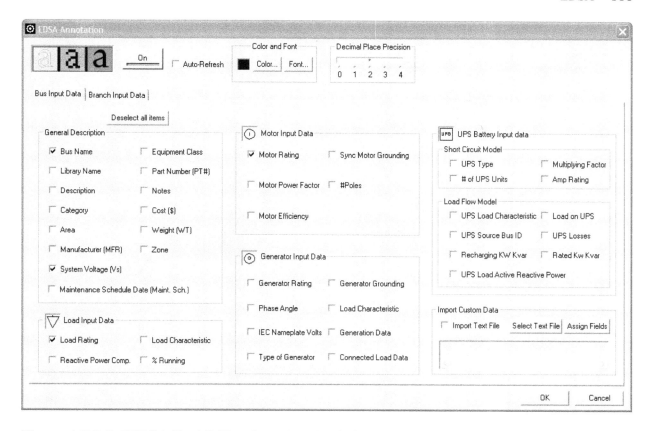

Figure 4-7-2-3 "EDSA Tools" "Bus Input Data" window.

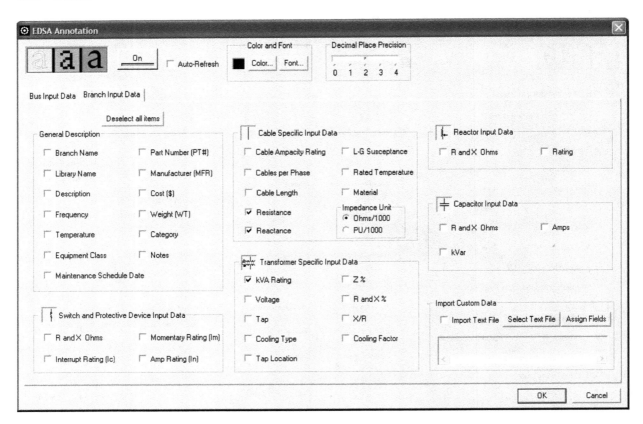

Figure 4-7-2-4 "EDSA Tools" "Branch Input Data" window.

9) The one-line diagram is in Figure 4-7-2-5.

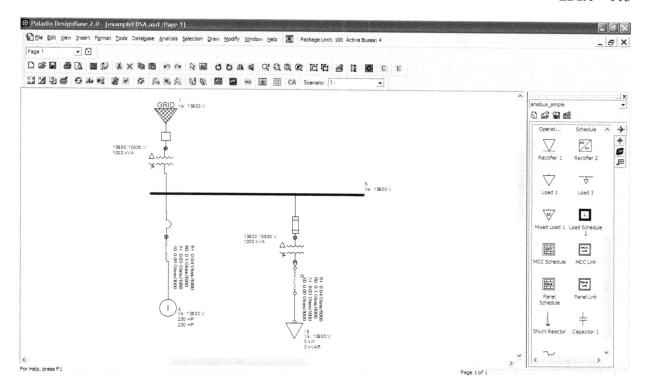

Figure 4-7-2-5 Work window of the Figure 1-3-1 "exampleEDSA" one-line diagram with default transformer winding connections and default voltages. No system data has been entered yet.

10) *Paladin DesignBase* has a "Make Straight" command to straighten one-line diagram leads. It is in the upper row of icons, fourth to the left from the toolbar right. Applying it to the circuit produces Figure 4-7-2-6.

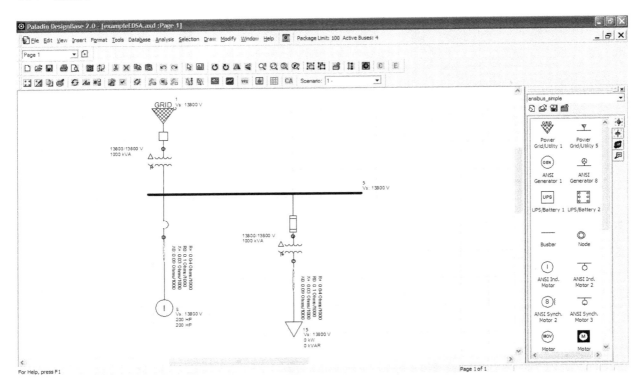

Figure 4-7-2-6 Work window of the Figure 4-7-2-5 after applying the "Make Straight" command.

4.7.3 POWER-FLOW ANALYSIS

4.7.3.1 Circuit Data Entry

Paladin DesignBase has many data windows for each component. With these, the user can input a great amount of data. However, most of this data is not needed for power-flow (load flow) and balanced short-circuit analyses. Here, only required data will be entered.

Underlined values are entered into *Paladin DesignBase*.

4.7.3.1.1 "Power Grid/Utility 1" data

1) Double left-click on the "Power Grid/Utility 1" icon. On the window that opens enter 3800 V rms, the value from the manual analysis Section 1.3.1.2.1 for the "System Volt". See Figure 4-7-3-1-1-1.

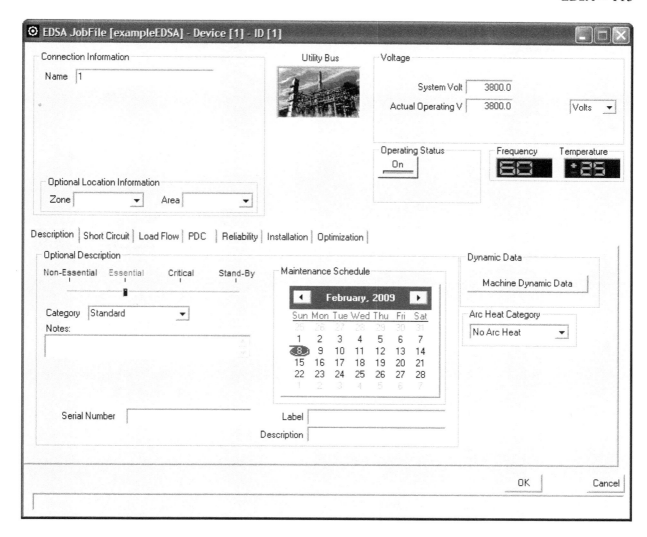

Figure 4-7-3-1-1-1 "Power Grid/Utility 1" first window.

2) On the same window, select the "Short Circuit" tab. Then in the appearing window in the "Power Data Type" select "Per Unit". The "Utility Base kVA', "3 Phase Per Unit Values", and "Grounding" will be available to receive values. For power-flow analysis and balanced short-circuit analysis, values for the negative and zero sequence impedances and connection configuration are not needed. Here, those impedances will be set to zero and the connection configuration will be left at its default value. They will be corrected later when analyzing a line-to-ground short circuit in Section 4.7.5.

Using values from Section 1.3.1.2.2 to enter the following data:

"Utility Base kVA" = Transformer T1 Base kVA = 600 kVA
"R +" = $RS1_{pu}$ = .008310
"X +" = $XS1_{pu}$ = .01246

3) Left-click "OK" after the data is entered. The filled out "Short Circuit" tab window is shown in Figure 4-7-3-1-1-2.

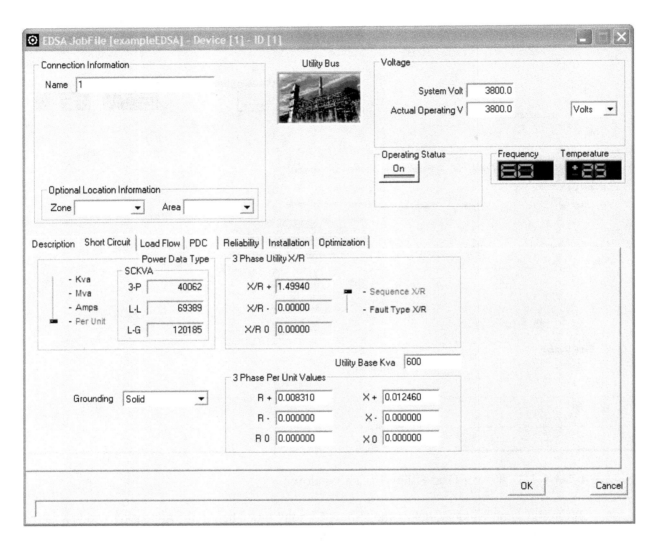

Figure 4-7-3-1-1-2 "Power Grid/Utility 1" "Short Circuit" tab.

4.7.3.1.2 Circuit breaker C1 and C2 and fuse F1 data

By default, *Paladin DesignBase* automatically enters small impedances for circuit breakers and fuses. The default C1, C2, and F1 impedances are much smaller than those of the cables and components that they are in series with so they have no significant impact on the circuit. No data input is necessary.

4.7.3.1.3 Transformer T1 data

1) *Paladin DesignBase* has a library of transformer data. A transformer's voltages are entered and then a typical transformer of the same kVA rating can be selected. *Paladin DesignBase* will automatically choose needed impedances. However, the example transformer T1 in Section 1.3.1.1.1 is not a typical transformer. Its impedances must be entered manually into the program.

2) At the top of the transformer window enter the transformer nominal voltages and power capacity. See Figure 4-7-3-1-3-1.

The entered values are:
"System Voltage" "From Voltage"= 3800
"System Voltage" "To Voltage" = 480
"Nameplate" "From Voltage"= 3800
"Nameplate" "To Voltage" = 480
"kVA Rating" = 600.00

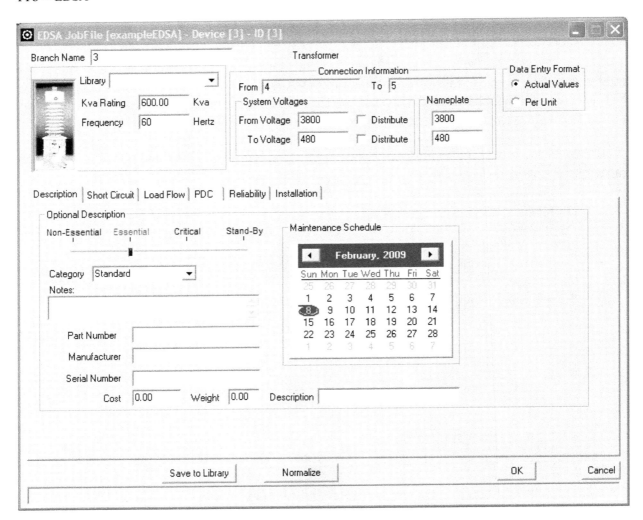

Figure 4-7-3-1-3-1 Transformer T1 first window of data.

3) On the "Short Circuit" tab window, enter the percent positive sequence impedances of Section 1.3.1.1.1. Zero sequence impedances will be set to zero. They are not needed for power-flow and balanced short-circuit calculations. Also, the transformer winding connections are not needed. They will be left at their default connections. Figure 4-7-3-1-3-2 shows the entered values.

The entered values are:
"%R +" = <u>1.73</u>
"%X +" = <u>1.52</u>

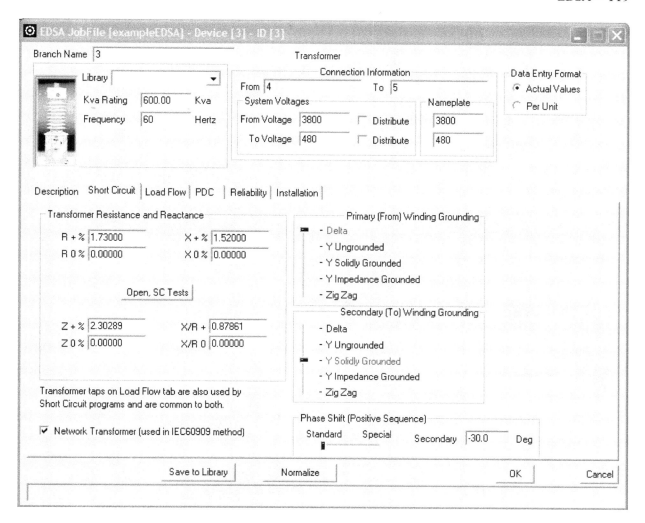

Figure 4-7-3-1-3-2 Transformer T1 "Short Circuit" tab window data.

4.7.3.1.4 Transformer T2 data

1) Transformer T2 data is taken from the manual analysis information in Section 1.3.1.1.2. Data is entered the same way as it was for transformer T1.

The entered values are:
"System Voltage" "From Voltage"= 480
"System Voltage" "To Voltage" = 120
"Nameplate" "From Voltage"= 480
"Nameplate" "To Voltage" = 120
"kVA Rating" = 90.00
"%R +" = 2.00
"%X +" = 1.80

4.7.3.1.5 Data for the cable from circuit breaker C2 to motor M

1) From the manual analysis information in Section 1.3.1.3.3:

"Cable Length" = <u>200</u> ft
"R + Ohms/1000 ft" = <u>0.078</u>, (Ω/1000 ft.)
"X + Ohms/1000 ft" = <u>0.030</u>, (Ω/1000 ft.)

2) Enter the length of the cable and select "User Defined" in the first cable window as shown in Figure 4-7-3-1-5-1.

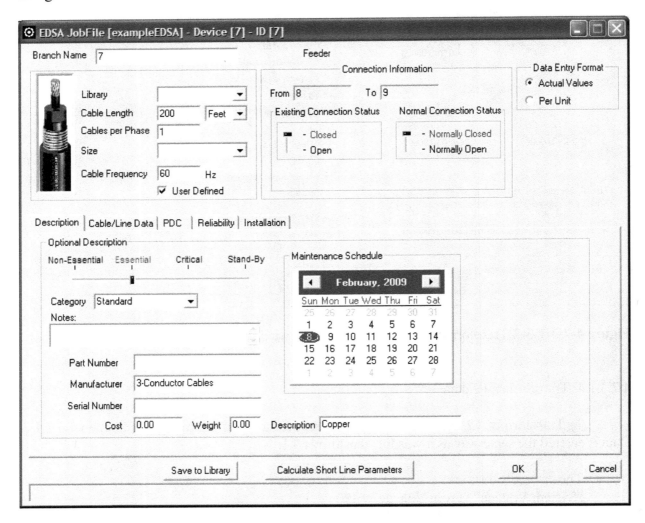

Figure 4-7-3-1-5-1 First cable window.

3) Open the "Cable/Line Data" tab window and enter the positive sequence impedances. Zero sequence impedances are not needed for power-flow and balanced short-circuit calculations. Here, zero will be entered for those impedances. Correct values will be entered later in the unbalanced short-circuit analysis section. The "Cable/Line Data" tab window is shown in Figure 4-7-3-1-5-2.

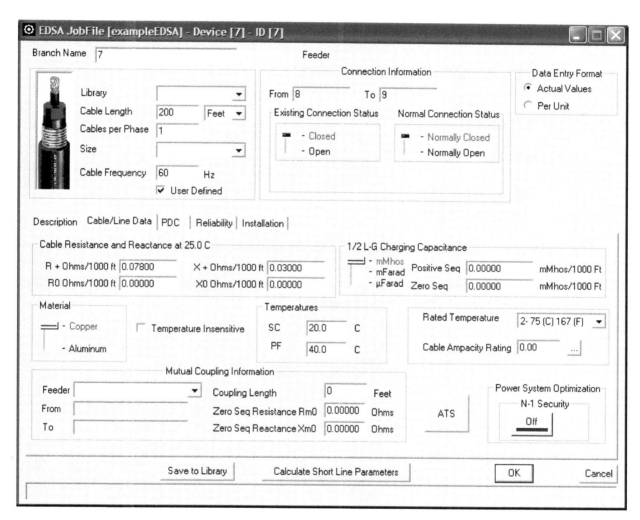

Figure 4-7-3-1-5-2 "Cable/Line Data" tab window.

4.7.3.1.6 Data for the cable from transformer T2 to static load R

1) From the manual analysis information in Section 1.3.1.3.4:

Length = <u>200</u> ft
Impedance (per conductor) Pos. R = <u>0.098</u>, (Ω/1000 ft.)
Impedance (per conductor) Pos. X = <u>0.030</u>, (Ω/1000 ft.)

2) The cable data is entered here as it was in 4.7.3.1.5.

4.7.3.1.7 Motor M data

1) Motor M data is taken from the manual analysis information in Section 1.3.1.4.

Ratings HP = <u>250</u>
Ratings kV = <u>0.48</u>

2) *Paladin DesignBase* has a motor in its library that fits these parameters. On the first motor window select "Library". The "Library" allows the selection of "250 HP-480 V". Once "250 HP-480 V" is selected and "OK" left-clicked, all needed motor data is entered. The first motor window, where "Library" is selected, is shown in Figure 4-7-3-1-7-1.

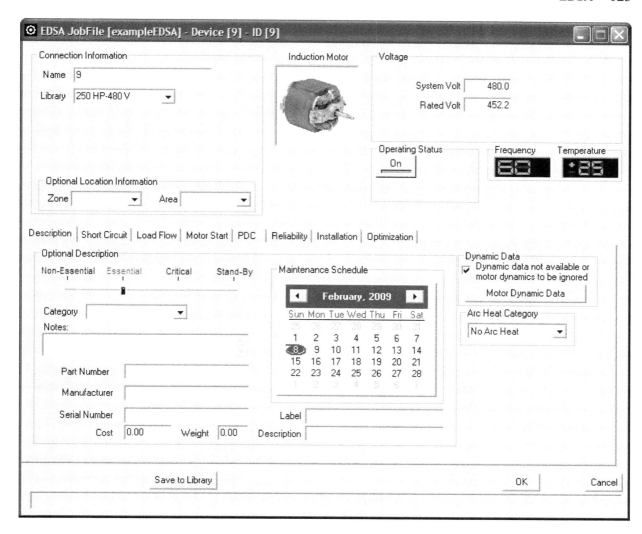

Figure 4-7-3-1-7-1 Motor M "Induction Machine Editor Mtr1" "Description" tab window.

3) The *Paladin DesignBase* library selects a motor efficiency and power factor that is close to, but different from, that used in the manual, *SKM*, and *ETAP* analyses. To make it easier to compare results, the efficiency and power factor in the *Paladin DesignBase* "Short Circuit" tab window will be over-written with the previously used values of:

"% Efficiency" = <u>92.86</u>
"% Power Factor" = <u>91.8</u>

4.7.3.1.8 "Static Load" R data

1) "Static Load" R data is taken from the manual analysis information in Section 1.3.1.5.

System Voltage = Rated Voltage = <u>120</u>
Electrical Rating kW = <u>30.00</u>
Electrical Rating kVAR = <u>0.00</u>
% Efficiency = <u>100.00</u>

2) For power-flow analysis and balanced short-circuit analysis, values for the negative and zero sequence impedances and connection configuration are not needed and can be left at default values.

3) Enter the voltage, power, and efficiency data into the "Load Flow" tab window. See Figure 4-7-3-1-8-1.

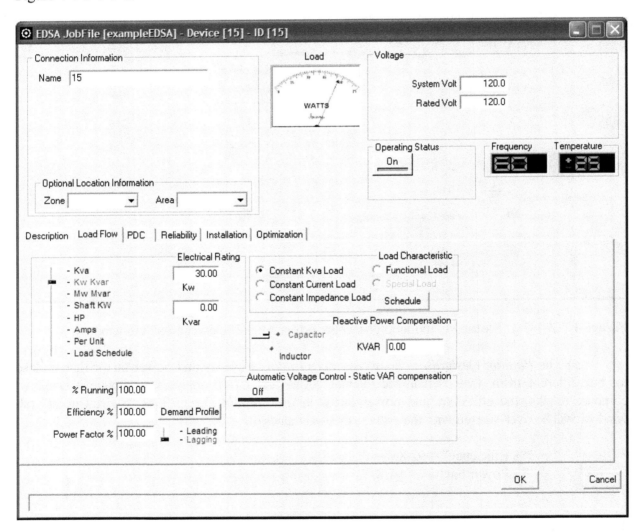

Figure 4-7-3-1-8-1 "Load 1" "Load Flow" tab window.

4.7.3.2 Power-flow Analysis

1) Left-click on the "Adv. Power Flow" icon. This is in the middle of the "EDSA tools" toolbar. It looks like two arrows at a 45 degree angle from the horizontal with a diamond-like shape next to them. This will cause an "Adv. Power Flow" toolbar to appear beneath the "Adv. Power Flow" icon. See Figure 4-7-3-2-1.

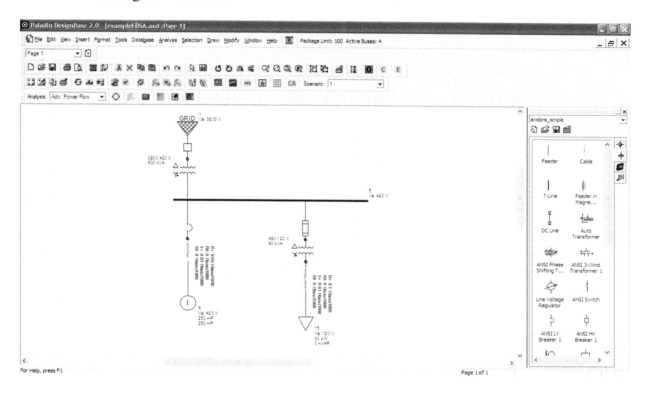

Figure 4-7-3-2-1 "Adv. Power Flow" toolbar in bottom row or toolbars.

2) Neaten the one-line diagram by turning off the display of the cable impedances. Do this by undoing the steps that were done in Step 8) of Section 4.7.2.

3) On the "Adv. Power Flow" toolbar left-click on its "Back Annotation" icon. This icon looks like an eye and is one icon to the left from the toolbar right. Note this "Back Annotation" is different from the "Back Annotation" that is in the toolbar above it. On the "Adv. Power Flow" "Back Annotation" window with the "AC Power Flow" tab, select "System Voltage (Vs)", "Bus Voltage", and "Current Flow". Set the "Bus Voltage" units to "V" and the "Current Flow" units to "A". Left-click "OK". See Figure 4-7-3-2-2.

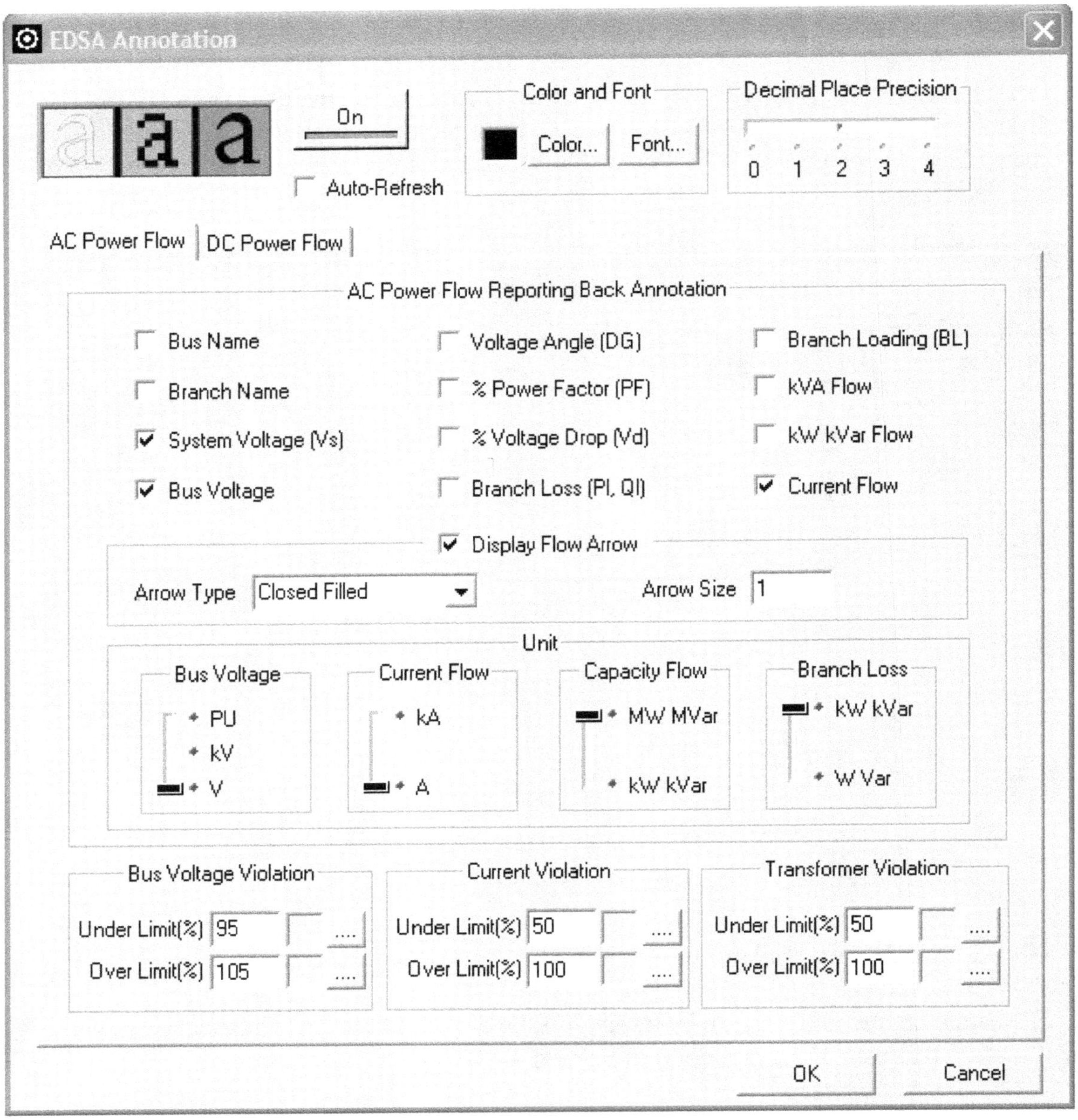

Figure 4-7-3-2-2 "Adv. Power Flow" "AC Power Flow" tab "Back Annotation" window.

4) On the "Adv. Power Flow" toolbar, left-click on its "Analyze" icon. This icon looks like a calculator and is the third icon from the toolbar left. This will cause a tabulated output page to appear. See Figure 4-7-3-2-3.

Exit Print Printer Font Screen Font Clipboard Save As DONE

```
                    EDSA Advanced Power Flow Program V6.00.00
                    =========================================

Project No.  :                        Page      : 1
Project Name:                         Date      : 02/15/2009
Title        :                        Time      : 00:53:26 pm
Drawing No.  :                        Company   :
Revision No.:                         Engineer  :
Jobfile Name: exampleEDSA             Check by  :
Scenario     :                        Date      :

Electrical One-Line 3-Phase to Single-Phase project


                                Bus Result
                                ==================
```

Bus Info & Sys kV			Voltage		Generation		Static Load		Motor Load		Load Flow Results				
Name	Type	KV	% Mag.	Ang.	MW	Mvar	MW	Mvar	MW	Mvar	To Bus Name	MW	Mvar	Amp	% PF
1	Swing	3.80	100.00	0.0	0.25	0.09	0.00	0.00	0.00	0.00	4	0.25	0.09	40	93.9
15	P_Load	0.12	93.71	-61.4	0.00	0.00	0.03	0.00	0.00	0.00	14	-0.03	0.00	155	100.0
5	Busbar	0.48	99.07	-30.2	0.00	0.00	0.00	0.00	0.00	0.00	12	0.03	0.00	39	100.0
											8	0.20	0.09	270	92.0
											4	-0.24	-0.09	39	93.9
9	P_Load	0.48	97.33	-30.2	0.00	0.00	0.00	0.00	0.21	0.09	8	-0.21	-0.09	275	92.2

Figure 4-7-3-2-3 "Adv. Power Flow" tabulated output data.

5) To see the currents and voltages displayed on the one-line diagram left-click on "DONE" on the tabulated output. This will return the screen to the one-line diagram, but this time the one-line diagram contains bus and branch currents and voltages. See Figure 4-7-3-2-4.

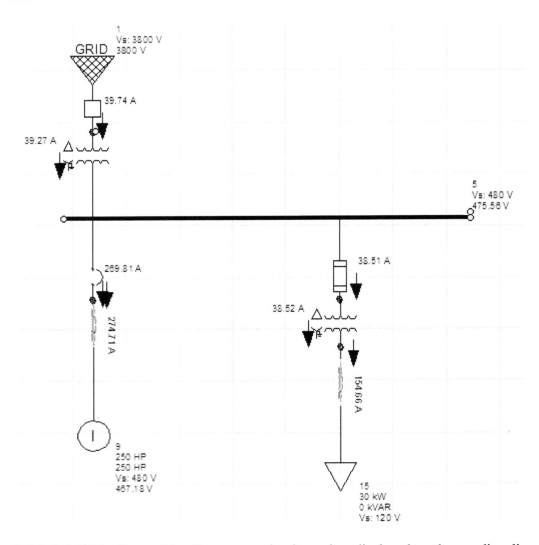

Figure 4-7-3-2-4 "Adv. Power Flow" current and voltage data displayed on the one-line diagram.

6) Notice that the currents shown in series components are not equal. For example, the current going through the motor circuit breaker is 269.81 A rms, but the current going through the cable that it is in series with it is 274.71 A rms. This is not correct, the currents should be equal. The reason this happens is that *EDSA* uses iterative numerical methods to solve the system. Numerical methods do not always produce an exact answer. They produce an answer within a certain tolerance.

7) To decrease the differences between the series currents in Figure 4-7-3-2-4, left-click on the "Options" icon in the "Adv. Power Flow" toolbar. This opens the "Options" window where calculation tolerances and numerical methods can be selected. In this window change the "Tolerance MVA" and "Tolerance %" to .001. Then left-click "OK". See Figure 4-7-3-2-5.

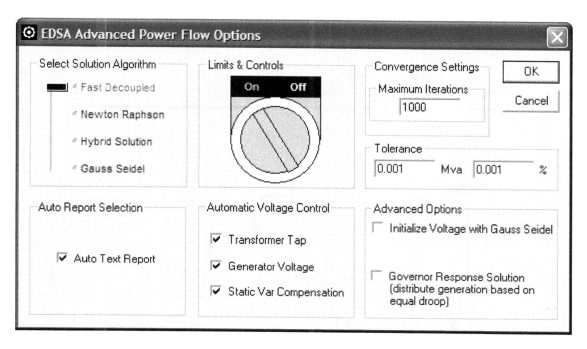

Figure 4-7-3-2-5 "Adv. Power Flow" "Options" window.

8) Left-click on "Analyze" again. The result is in Figure 4-7-3-2-6. Notice that the series currents are much closer now.

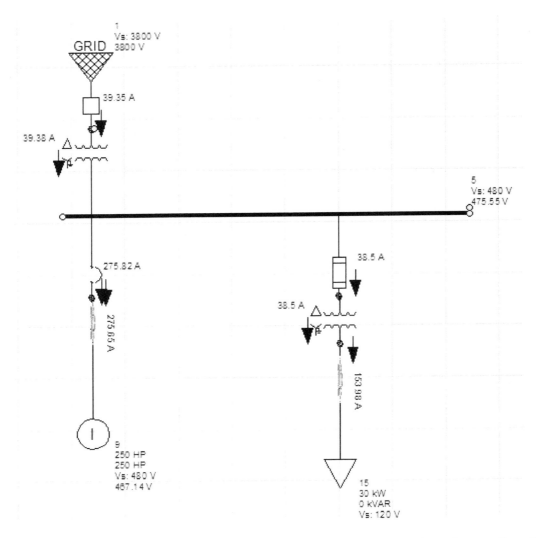

Figure 4-7-3-2-6 "Adv. Power Flow" current and voltage data displayed on the one-line diagram after decreasing the calculation error tolerance.

4.7.4 BALANCED SHORT-CIRCUIT ANALYSIS

1) Left-click on the "AC Short Circuit" icon. This is near the middle of the "EDSA tools" toolbar. It looks like lightning bolt. This will cause an "AC Short Circuit Tools" toolbar to appear beneath the "AC Short Circuit" icon. On the "AC Short Circuit Tools" toolbar select the "Analysis:" type to be "AC ANSI/IEEE". See Figure 4-7-4-1.

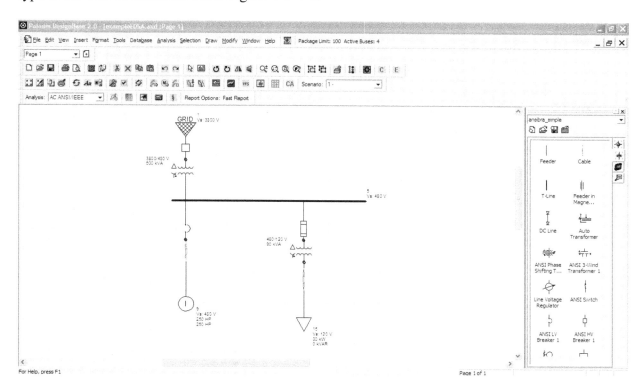

Figure 4-7-4-1 Window showing "AC Short Circuit Tools" toolbar with the "AC ANSI/IEEE" selection.

2) Left-click on the "AC Short Circuit Tools" toolbar "Options" icon. This icon looks like a lightning bolt and is on the left of the "AC Short Circuit Tools" toolbar. On the window that appears, select "All Buses" and left-click "OK". See Figure 4-7-4-2.

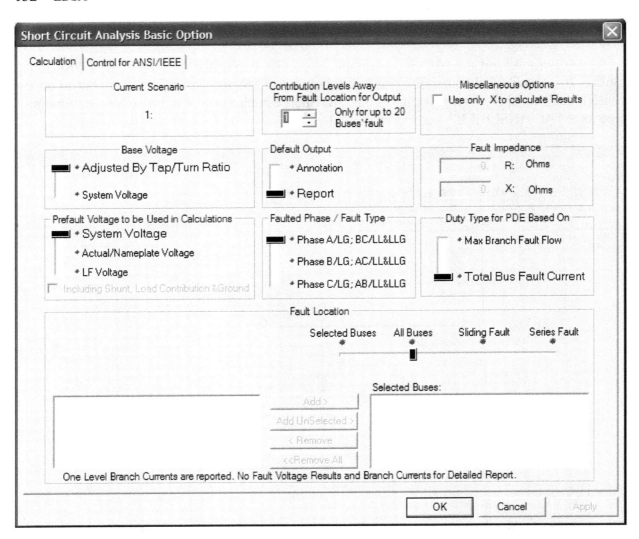

Figure 4-7-4-2 Window showing "AC Short Circuit Tools" toolbar "Calculation" "Options" window.

3) Left-click on the "AC Short Circuit Tools" toolbar "Back Annotation" icon. This icon looks like an eye and is in the middle of the "AC Short Circuit Tools" toolbar. Note this is not the same "Back Annotation" as was used before. On the window that opens left-click "3 Phase" "Fault Type", select a "5 Cycle" "Fault Time" and left-click "OK". See Figure 4-7-4-3.

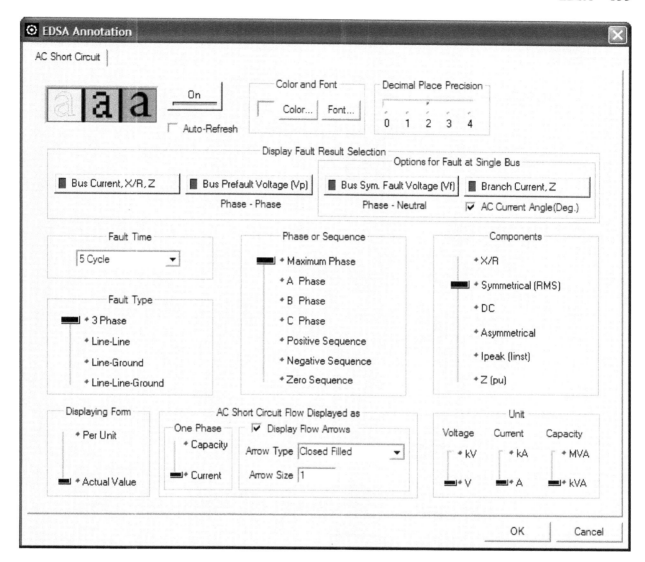

Figure 4-7-4-3 Window showing "AC Short Circuit Tools" toolbar "Back Annotation" window for a "3 Phase" short circuit.

4) Left-click on the "AC Short Circuit Tools" toolbar "Analyze" icon. This icon looks like a calculator and is the fourth icon to the right. It produces a tabulated data report. The balanced three-phase short-circuit currents are at the bottom. Each short-circuit current was calculated assuming one bus short circuit at a time. See Figure 4-7-4-4.

```
                                 EDSA

                   3-Phase Short Circuit v6.50.00

Project No. :                         Page    : 1
Project Name:                         Date    : 02/22/2009
Title       :                         Time    : 03:07:08 pm
Drawing No. :                         Company :
Revision No.:                         Engineer:
Jobfile Name: exampleEDSA             Check by:
Scenario    : 1 :                     Date    :
------------------------------------------------------------------
        Electrical One-Line 3-Phase to Single-Phase project


                        ------------------
                          System Summary
                        ------------------


Base MVA                          :   100.000
System Frequence(Hz)              :   60

# of Total Areas Named            :   0
# of Total Zones Named            :   0
# of Total Buses                  :   8
# of Active Buses                 :   8
# of Total Branches               :   7

# of Active Sources               :   1
# of Active Motors                :   1
# of Active Shunts                :   1
# of Transformers                 :   2
Reference Temperature(°C)         :   20.0
Impedance Displaying Temperature(°C) : 25.0




                        --------------------
                          Calculation Options
                        --------------------


Calculating All or Mult-Buses Fault with Fault Z =  0.00000  + j  0.00000 Ohms

Fault Phases:
    Phase A for Line-Ground Fault
    Phase B,C for Line-Line or Line-Line-Ground Fault

ANSI/IEEE Calculation:
    Using ANSI Std. C37.010-1979 or above.
    Separate R and X for X/R, Complex Z for Fault Current
    The Multiplying Factors to calculate Asym and Peak are Based on Actual X/R
    Peak Time Applies ATPC Equation

Transformer Phase Shift is not considered.
Generator and Motor X/R is constant.
Base     Voltages  : Adjusted by Tap/Turn Ratio
Prefault Voltages  : Use System Voltages
------------------------------------------------------------------


Jobfile Name: exampleEDSA                 Page    : 2


                 ---------------------------------------
                 Bus Results:  5 Cycle--Symmetrical
                 ---------------------------------------


                                  Thevenin Imped.  ANSI
                      Pre-Flt 3P Flt. ----------------- ------
Bus Name                 V       A    Z+(pu)   Zo(pu)  3P X/R
------------------------ ------- ----- -------- -------- ------
1                        3800    6151  2.4701   2.4962  1.5730
15                        120    2779  173.157  307.106 0.4172
5                         480   19614  6.1324   3.8381  1.3003
9                         480    9557  12.5855  17.9514 1.3290
```

Figure 4-7-4-4 "AC Short Circuit" "3 Phase" tabulated output data. Balanced three-phase short-circuit currents are at the bottom.

4.7.5 UNBALANCED SHORT-CIRCUIT ANALYSIS

Unbalanced short-circuit analysis requires more information than power-flow and balanced short-circuit analysis require. For power-flow and balanced short-circuit analyses only *symmetrical components* positive sequence impedances are needed. For unbalanced short-circuit analysis *symmetrical components* negative and zero sequence impedances are also needed. Furthermore, unlike balanced analyses, correct transformer connections are needed (i.e. Y-Y, Y-Y(ground), Y-Δ, etc.).

4.7.5.1 "Power Grid/Utility 1" Data

1) Double left-click on the circuit's "Power Grid/Utility 1" icon again and left-click on the "Short Circuit" tab. Using values from the manual analysis Section 1.3.1.2.2, enter the values:

"R -" = $RS2_{pu}$ = .008310
"X -" = $XS2_{pu}$ = .01246
"R 0" = $RS0_{pu}$ = .004155
"X 0" = $XS0_{pu}$ = .006232

The window is shown in Figure 4-7-5-1-1.

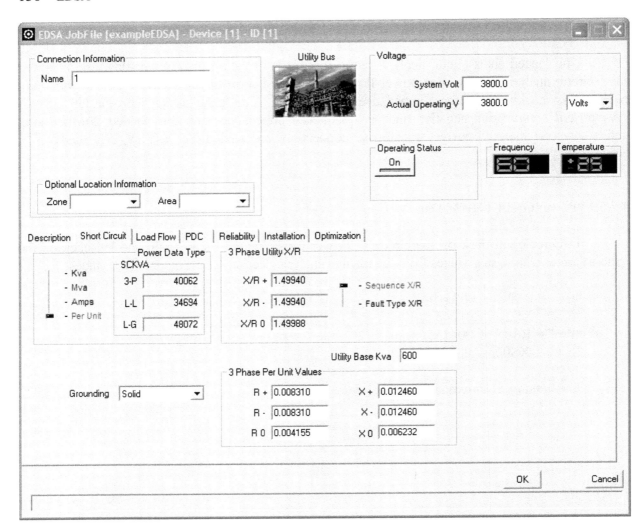

Figure 4-7-5-1-1 "Power Grid/Utility 1" "Short Circuit" tab with correct negative and zero sequence impedances.

4.7.5.2 Circuit Breaker C1 and C2 and Fuse F1 Data

No data input is necessary.

4.7.5.3 Transformer T1 Data

1) *Paladin DesignBase* assumes that transformer negative sequence impedances are the same as positive sequence impedances.

2) Double left-click on the circuit's transformer T1 "ANSI Transformer 1" icon again and left-click on the "Short Circuit" tab. Change the "Primary (From) Winding Grounding" to "Y Solidly Grounded" and add values from the manual analysis Section 1.3.1.1.1:

"R 0 %" = $RT10_{pu}$ = <u>1.73</u>
"X 0 %" = $XT10_{pu}$ = <u>1.52</u>

The window is shown in Figure 4-7-5-3-1.

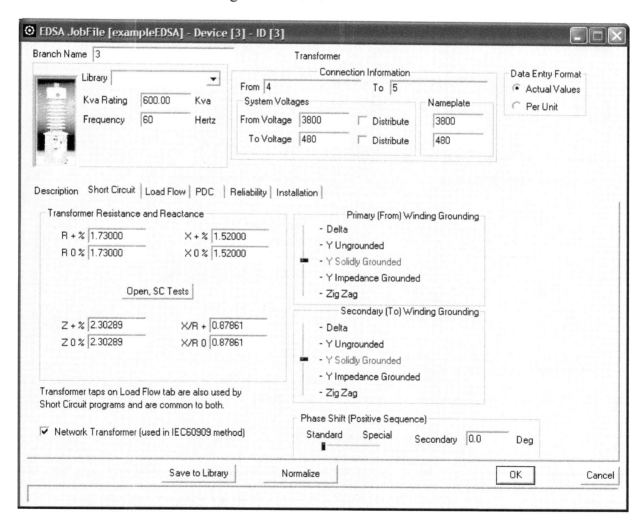

Figure 4-7-5-3-1 Transformer T1 "ANSI Transformer 1" "Short Circuit" tab with correct zero sequence impedances and primary winding connection. (Note: "ANSI Transformer 1" refers to the type of transformer, not the transformer T1.)

4.7.5.4 Transformer T2 Data

1) The transformer T2 data needs to changed, just as it was for the transformer T1. Change the "Primary (From) Winding Grounding" to "Y Ungrounded" and add values from the manual analysis Section 1.3.1.1.2:

"R 0 %" = RT20$_{pu}$ = <u>1.00</u>
"X 0 %" = XT20$_{pu}$ = <u>7.60</u>

The window is shown in Figure 4-7-5-4-1.

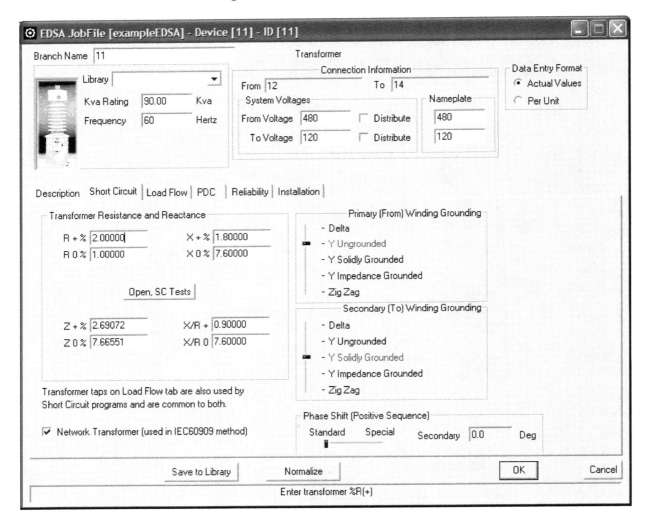

Figure 4-7-5-4-1 Transformer T2 "ANSI Transformer 1" "Short Circuit" tab with correct zero sequence impedances and primary winding connection. (Note: "ANSI Transformer 1" refers to the type of transformer, not the transformer T1.)

4.7.5.5 Data for the Cable from Circuit Breaker C2 to Motor M

Paladin DesignBase assumes that the negative sequence impedances are the same as the positive. The zero sequence impedances need to be entered manually. In Section 1.3.1.3.3 the zero sequence impedances are the same as the positive. They should be entered into the circuit breaker C2 to motor M "Cable" "Cable/Line Data" tab window as shown in Figure 4-7-5-5-1.

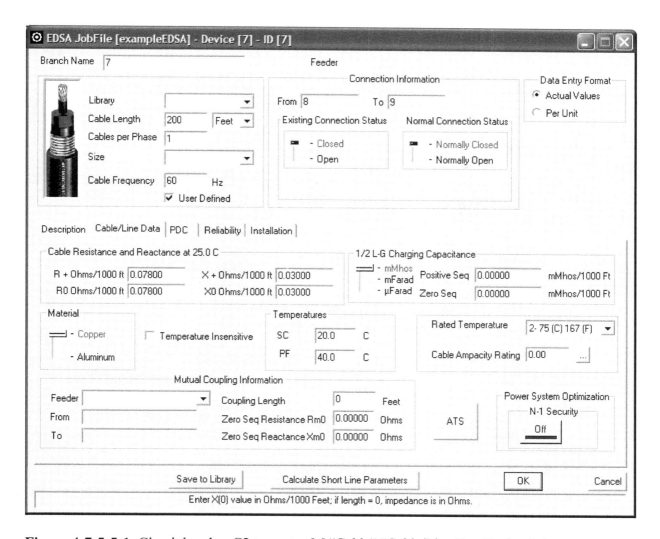

Figure 4-7-5-5-1 Circuit breaker C2 to motor M "Cable" "Cable/Line Data" tab window.

4.7.5.6 Data for the Cable from Transformer T2 to Static Load R

Paladin DesignBase assumes that the negative sequence impedances are the same as the positive. The zero sequence impedances need to be entered manually. In Section 1.3.1.3.4 the zero sequence impedances are the same as the positive. They should be entered into the transformer T2 to static load R "Cable" "Cable/Line Data" tab window as they were in Section 4.7.5.5.

4.7.5.7 Motor M Data

Paladin DesignBase selected typical negative and zero sequence impedances when motor data was input from the *Paladin DesignBase* library. It is not necessary to add further data.

4.7.5.8 Static Load R Data

Paladin DesignBase assumes that the negative sequence impedances are the same as the positive. The zero sequence impedances are infinite, since the neutral is isolated. It is not necessary to add further data to the *Paladin DesignBase* program.

4.7.5.9 Unbalanced Short-Circuit Analysis of a Line-to-Ground Short Circuit

The procedure is practically the same as that used for balanced short-circuit analysis in Section 4.7.4.

1) As in Section 4.7.4 left-click on the "AC Short Circuit" icon. Then on the "AC Short Circuit Tools" toolbar, select the "Analysis:" type to be "AC ANSI/IEEE". See Figure 4-7-4-1.

2) As in Section 4.7.4 left-click on the "AC Short Circuit Tools" toolbar "Options" icon. On the window that appears, select "All Buses" and left-click "OK". See Figure 4-7-4-2.

3) Left-click on the "AC Short Circuit Tools" toolbar "Back Annotation" icon. On the window that opens, left-click "Line-Ground", "5 Cycle", and "Maximum Phase". Then, left-click "OK". See Figure 4-7-5-9-1.

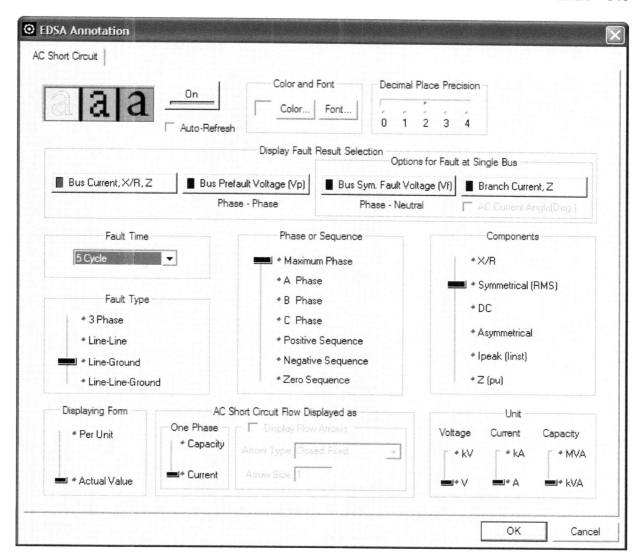

Figure 4-7-5-9-1 Window showing "AC Short Circuit Tools" toolbar "Back Annotation" window for a "Line-Ground" short circuit.

4) Left-click on the "AC Short Circuit Tools" toolbar "Analyze" icon. This produces a tabulated data report. The line-to-ground short circuit currents are at the bottom of the report. Each short-circuit current was calculated assuming one bus short circuit at a time. See Figure 4-7-5-9-2.

```
                                      EDSA

                       3-Phase Short Circuit v6.50.00

Project No. :                              Page    : 1
Project Name:                              Date    : 02/22/2009
Title     :                                Time    : 02:18:54 pm
Drawing No. :                              Company :
Revision No.:                              Engineer:
Jobfile Name: exampleEDSA                  Check by:
Scenario   : 1 :                           Date    :
_____
         Electrical One-Line 3-Phase to Single-Phase project


                            _____
                            System Summary
                            _____

Base MVA                              :  100.000
System Frequence(Hz)                  :  60

# of Total Areas Named                :  0
# of Total Zones Named                :  0
# of Total Buses                      :  8
# of Active Buses                     :  8
# of Total Branches                   :  7

# of Active Sources                   :  1
# of Active Motors                    :  1
# of Active Shunts                    :  1
# of Transformers                     :  2
Reference Temperature(°C)             :  20.0
Impedance Displaying Temperature(°C)  :  25.0



                         _____
                         Calculation Options
                         _____


Calculating All or Mult-Buses Fault with Fault Z =  0.00000  + j  0.00000 Ohms

Fault Phases:
     Phase A for Line-Ground Fault
     Phase B,C for Line-Line or Line-Line-Ground Fault

ANSI/IEEE Calculation:
     Using ANSI Std. C37.010-1979 or above.
     Separate R and X for X/R, Complex Z for Fault Current
     The Multiplying Factors to calculate Asym and Peak are Based on Actual X/R
     Peak Time Applies ATPC Equation

Transformer Phase Shift is not considered.
Generator and Motor X/R is constant.
Base     Voltages  : Adjusted by Tap/Turn Ratio
Prefault Voltages  : Use System Voltages
_____


Jobfile Name: exampleEDSA                      Page    : 2


                  _____
                  Bus Results:  5 Cycle--Symmetrical
                  _____
```

			Thevenin Imped.		ANSI
	Pre-Flt	LG Flt.	---------------		------
Bus Name	V	A	Z+(pu)	Zo(pu)	3P X/R
1	3800	7365	2.4701	1.2484	1.5730
15	120	0	173.158	—	0.4172
5	480	22438	6.1324	3.8381	1.3003
9	480	10037	12.5855	10.8498	1.3290

Figure 4-7-5-9-2 "AC Short Circuit" "Line-Ground" tabulated output data. Line-to-ground short-circuit currents are at the bottom.

4.8 *PALADIN DESIGNBASE* REFERENCES

1) *EDSA, Paladin DesignBase 2.0 Documents*, 2008. This is a CD containing text tutorials and manuals. It is free and comes with the Trial Version CD. A person learning *Paladin DesignBase* should read them.

2) *Paladin DesignBase 2.0 Tutorials*, 2008. This is a CD that plays tutorial videos on *Windows*. It is free and comes with the Trial Version CD. Its tutorials can also be seen on the *EDSA* website. A person learning *Paladin DesignBase* should see them.

5.0 MISCELLANEOUS

5.1 POWER SYSTEM SIMULATION SOFTWARE DATA TIPS

The computer programmer's saying, "Garbage in, garbage out", applies to power system simulation software. Be very careful when setting up power system models and entering data.

One of the simulation software companies mentioned above states that occasionally an impedance of zero will cause a divide by zero error in their software. This problem did not occur in this book's example analyses, where zero values were used. However, if a divide by zero error occurred during the analysis of a large 1000+ bus systems it could be very aggravating. The company therefore recommends entering non-zero values for all sequence impedances. They recommend non-zero values for negative and zero sequence impedances even for load flow and balanced short-circuit studies.

In the example load flow and balanced short-circuit studies, proper zero sequence and negative sequence impedances and transformer connections (Y-Y, Y-Y(ground), Y-Δ, etc.)-were not entered. This did not cause difficulties. However, in an industrial study this could lead to a problem. In an industrial study, at some unknown future date, someone else might mistakenly use the same data to do an unbalanced short-circuit study. If possible, enter correct impedances for all sequences when originally entering data.

5.2 DIFFERENT RESULTS FROM DIFFERENT ANALYSES

The *ETAP*, *SKM*, and *Paladin DesignBase* produced output data within a +/-2% variation. The manual analysis output data differed from the programs' averaged output data by up to 7%.

The major reason for the differences between the output data of the manual analysis and the programs is the motor modeling method. The manual analysis used a fixed resistance in series with a fixed inductive impedance to model the motor. *SKM*, *ETAP*, and *Paladin DesignBase* use a more accurate motor model that effectively makes motor impedance decrease as applied voltage decreases. The result was that the manual analysis predicted lower motor currents.

Another reason for differences in the output data is the way the circuits were solved. The manual analysis used phasor and *symmetrical components* methods to solve the system's equations analytically. *SKM*, *ETAP*, and *Paladin DesignBase* used numerical analysis methods. Numerical analysis methods use approximation techniques to solve the equations. The different approximation techniques may introduce significant errors. An example of a numerical analysis error can be seen in Section 4.7.3.2 with the *Paladin DesignBase* power-flow analysis. In the first results of that analysis, two series currents that should have been the same were significantly different. See Figure 4-7-3-2-4. This was fixed by decreasing the MVA tolerance. See Figure 4-7-3-2-6.

5.3 COMMON POWER SYSTEM ANALYSIS STEPS

Regardless of the computer program used, there are certain steps that must be taken to carry out a power system analysis.

1) Draw the circuit diagram.
　　This may require existing or proposed circuit drawings, a site visit, and meetings with the customer.
2) Determine values for the circuit components and cabling.
　　This may require meetings with the customer, drawings, a site visit with measurements, viewing of old drawings, viewing of equipment specification sheets, and contact with equipment manufacturers. This is often the most difficult step. Remember to document the rationale behind your choices.
3) Convert the data to a format that can be accepted by the program.
4) Enter the data into the program.
5) Run the program.
6) Check the results.
7) Print the results.

5.4 WHAT DOES THE SOFTWARE COST?

Software prices vary greatly. For under $2,000, one can purchase a single user license for a base module that, without other modules, does little more than create one-line diagrams. For about $60,000, one can purchase a nuclear power plant certified single user license for a program that includes many modules. Prices may be higher for programs that can handle larger power systems. The software companies may offer quantity discounts or other discounts. Prices may change with time and as new program versions are released. After the initial purchase there are fees for optional maintenance contracts and, after an initial startup period, technical support contracts.

Tuition for the software's classroom training ranges from $300 to $1,000 per day. $400 per day is typical.

Annual support and maintenance for the programs after the first year is about $1,000.

To get actual prices, the potential customer should determine what he needs and then request quotes.

5.5 OTHER POWER SYSTEM ANALYSIS PROGRAMS

There are many other power system analysis programs. Below is a partial list:

Company Name	Website
Amtech	http://www.amtech-power.co.uk/
ASPEN	http://www.aspeninc.com/
CAPE/Electrocon	http://www.electrocon.com/
CYME	http://www.cyme.com/
DataShare	http://www.datashare.com.au/
DIgSILENT	http://www.digsilent.de/
EasyPower	http://www.easypower.com/
EDR	http://www.edreference.com/default.asp
MilSoft	http://www.milsoft.com/
Neplan	http://www.neplan.ch/sites/en/default.asp
PSI	http://members.shaw.ca/powersoftech/

5.6 SPICE

Spice is an acronym for **S**imulation **P**rogram with **I**ntegrated **C**ircuit **E**mphasis. There are number of Spice programs available. The intended purpose of the Spice programs is to solve low power electronic circuits. However, they will also accurately solve high power circuits, such as those of power systems. In some cases, such as when unusual voltages are applied to power circuits, the Spice programs can do a better analysis than achievable with power system analysis programs.

Pspice is the most popular of the Spice programs. Details on applying *PSpice* to power systems can be found in the book, *PSpice Power Electronic and Power Circuit Simulation*, S. Tubbs, 2008, ISBN 978-0-9659446-9-4.

www.ingramcontent.com/pod-product-compliance
Lightning Source LLC
Chambersburg PA
CBHW080420060326
40689CB00019B/4311